Communicating Public Health Risk

This important volume provides not only an in-depth analysis of those risk communication strategies currently used to inform and educate the public about key health issues but also the risks and effects of radon, a natural but carcinogenic gas that so far has seen relatively little wider coverage.

As the leading cause of lung cancer worldwide after smoking, radon is an important yet hidden public health issue, but informing and educating the public about its hazards and dangers is far from straightforward. As well as offering a detailed overview of issues around radon itself, the book asserts that public health communication should be dialogic and interactive, culturally tailored to specific populations to ensure people comprehend and appreciate the risk to themselves and their environments. The challenges are, of course, significant in a pluralistic media landscape where disinformation and misinformation threaten the integrity of any message sent.

Featuring chapters from researchers across a range of disciplines, this enlightening book will interest students, scholars and professionals working in Public Health, Environment Health and Communication Studies.

José Sixto-García is Professor of Journalism at the Department of Communication Sciences of the University of Santiago de Compostela (Spain). He has a PhD in Communication and Journalism. He was the Director of the Social Media Institute (2013–2019), and his research is focused on new media, new narratives and social networks.

Sara Pérez-Seijo has a PhD in Communication, is Assistant Professor at the Universidade de Santiago de Compostela (Spain) and a member of Novos Medios research group. She was a visiting scholar at Universidade Nova de Lisboa, EUniWell's Communications Manager at the Universidade de Santiago de Compostela and Section Editor at *Profesional de la Información*.

Berta García-Orosa is Full Professor at the University of Santiago de Compostela (Spain). She holds a B.A. in Communication Sciences, a B.A. in Political and Administration Sciences and a PhD in Communication Sciences from the University of Santiago de Compostela. She has studied communication and politics for more than 20 years, collaborated in more than 50 research projects and published over 100 articles and chapters.

Communicating Public Health Risk

The Case of Radon Gas

Edited by José Sixto-García,
Sara Pérez-Seijo and
Berta García-Orosa

Routledge
Taylor & Francis Group

LONDON AND NEW YORK

First published 2025
by Routledge
4 Park Square, Milton Park, Abingdon, Oxon OX14 4RN

and by Routledge
605 Third Avenue, New York, NY 10158

Routledge is an imprint of the Taylor & Francis Group, an informa business

British Library Cataloguing-in-Publication Data
A catalogue record for this book is available from the British Library

Library of Congress Cataloging-in-Publication Data
Names: Sixto-García, José, editor. | Pérez-Seijo, Sara, editor. |
    García Orosa, Berta, editor.
Title: Communicating public health risk : the case of radon gas / edited by
    José Sixto-García, Sara Pérez-Seijo, and Berta García-Orosa.
Description: Abingdon, Oxon ; New York, NY : Routledge, 2025. |
    Includes bibliographical references and index.
Identifiers: LCCN 2024015390 (print) | LCCN 2024015391 (ebook) |
Subjects: MESH: Radon—toxicity | Air Pollutants, Radioactive—toxicity |
    Carcinogens, Environmental—toxicity | Risk Management—methods | Health
    Communication—methods | Communications Media
Classification: LCC TD885.5.R33 C67 2025 (print) | LCC TD885.5.R33
    (ebook) | NLM WN 615 | DDC 363.738/7014—dc23/eng/20240416
LC record available at https://lccn.loc.gov/2024015390
LC ebook record available at https://lccn.loc.gov/2024015391

ISBN: 978-1-032-61811-1 (hbk)
ISBN: 978-1-032-61813-5 (pbk)
ISBN: 978-1-032-61818-0 (ebk)

DOI: 10.4324/9781032618180

Typeset in Times New Roman
by Apex CoVantage, LLC

Contents

Figures

Tables

Contributors

Sylvain Andresz holds two degrees in nuclear engineering. He works as Senior Researcher at the Nuclear Protection Evaluation Centre (CEPN, France) on the evaluation and management of radiological risks under different exposure situations. He has ten years' experience in supporting radon management initiatives set by local territories in France. He was the chair of the "ALARA for Radon At Work" (A-RAW) Working Group of the European ALARA Network and the Secretary of ICRP Publication 126.

Rita Araújo, PhD in Communication, is Auxiliary Researcher at Communication and Society Research Centre, Universidade do Minho (Portugal). She was a Visiting Scholar at Hunter College, City University School of Public Health, New York. Her research interests include journalism, health communication, news sources, and health literacy.

Juan Miguel Barros Dios is Professor of Epidemiology and Public Health at the University of Santiago de Compostela (Spain), working in the Department of Preventive Medicine and Public Health. Affiliated with Consortium for Biomedical Research in Epidemiology and Public Health (CIBER en epidemiología y Salud Pública/CIBERESP) and the Health Research Institute of Santiago de Compostela (Instituto de Investigación Sanitaria de Santiago de Compostela—IDIS).

Pavel Sidorenko Bautista, PhD from the Faculty of Communication of the University of Castilla-La Mancha, is Professor and Researcher at the Faculty of Business and Communication of the International University of La Rioja (Spain). Member of the Research Group Procomm UNIR and co-responsible for the International Congress of New Narratives in the Digital Society 'Digit-ALL'.

Gabriela Coronel-Salas is Professor at Universidad Técnica Particular de Loja (Ecuador) and social Communicator. She holds a PhD in Communication and Journalism from the Universidade de Santiago de Compostela (Spain) under the research line of journalism and new technologies and Master in Science and Technology Studies from the Universidad de Salamanca (Spain). Manager of technologies focused on online communication and public communication of science. General editor of *Soy+Familia* magazine.

Carmen Costa-Sánchez is a professor and researcher in the Department of Sociology and Communication Science at the University of A Coruña (Spain). She is the Director of the Chair of Social Commitment, Communication and Corporate Reputation at the University of A Coruña. Her main lines of research are corporate communication, health communication and social media.

Alison Dowdall has an MSc in Instrumental Analysis from Dublin City University, Ireland. Her role as a scientist with the Environmental Protection Agency is to manage the delivery of EPA actions under the Irish Government's National Radon Control Strategy. Additionally, she is responsible for research with a particular focus on establishing metrics to assess the Strategy's progress over time.

Pablo Escandón-Montenegro holds a PhD in Contemporary Communication and Information from the Universidade Santiago de Compostela, Spain. He is currently the academic coordinator of the Master's Degree in Transmedia Communication at the Universidad Andina Simón Bolívar, Ecuador. His research focus is on digital platforms and narratives.

Tania Forja-Pena is a PhD student in Communication and Contemporary Information at University of Santiago de Compostela (USC) and a member of the Novos Medios research group at the same university. Her research is linked to risk communication, digital journalism, and disinformation through the analysis of media and social media.

Marta García-Talavera is Head of the Natural Radiation Section at CSN (Spain's Nuclear Safety Council). Chair of the IAEA ENVIRONET-NORM Project. Her research interests include public policy analysis, radiation protection, radon mapping, NORM management, sustainability in waste management, and contaminated land characterization and remediation.

Sofia Gomes holds a PhD (2019) in Communication Sciences from the UMinho (Portugal) on Health Journalism. She is a researcher at the Communication and Society Research Centre and her main research interests are in the areas of journalism, health journalism and sources of information.

Xosé López-García is Professor of Journalism at the Department of Communication Sciences at the University of Santiago de Compostela (Spain). Since 1994, he has coordinated the Novos Medios research group (GI-1641 NM). He is the principal investigator of the project Digital-native media in Spain: Strategies, competencies, social involvement and (re)definition of practices in journalistic production and diffusion (PID2021–122534OB-C21), funded by MCIN/AEI/10.13039/501100011033/and by "ERDF A way of making Europe."

Lila Luchessi has a PhD in Political Science and a Graduate in Social Communication Sciences. She is a full professor at the National University of Río

Negro and an Associate at the University of Buenos Aires (Argentina). She is the Director of the Institute for Research in Public Policies and Government of the National University of Río Negro and is a member of the academic committee of the Latin American Chair of Narratives Transmedia.

Meritxell Martell has a PhD in Environmental Sciences from the University of East Anglia, United Kingdom. She is the founder and director of the strategic environmental consultancy MERIENCE. She has over 20 years of experience as a consultant on risk communication and stakeholder engagement in complex environmental decision-making processes for different European and international organisations.

Lucía Martín de Bernardo Gisbert is a Predoctoral Researcher in Epidemiology and Public Health at the University of Santiago de Compostela (Spain). She is affiliated to the Department of Preventive Medicine and Public Health and the Cross-disciplinary Research Center in Environmental Technologies (CRETUS).

Leo McKittrick has an MSc in Instrumental Analysis from Dublin City University, Ireland. He is a scientist with the Environmental Protection Agency, where he manages the Citizen Science programme. His focus is on developing public participation programmes to increase awareness of environmental issues, such as the EPA's Clean Air Together project and the library loan scheme of radon digital monitors in Ireland.

Mariola Moreno Calvo has a PhD in Information and Communication Sciences from the University of Toulouse (France). She is a researcher and lecturer in communication studies at the International University of La Rioja (Spain) and a Member of the Research Group Procomm UNIR.

María-Cruz Negreira-Rey has a PhD in Communication, Universidade de Santiago de Compostela (Spain) and is now Assistant Professor of Journalism at the Department of Communication Sciences and member of Novos Medios research group. Her research focus is on local journalism and the development of local and hyperlocal media, digital journalism and social media.

Veronika Oláhné Groma is a scientific researcher working at the Centre for Energy Research in Hungary. She possesses a distinct passion for investigating population—especially students—engagement in her field of research, which primarily focuses on environmental physics, air quality analysis, and health risk assessment, in which she has nearly 20 years of experience.

Noel Pascual-Presa is a PhD student in Communication at the University of Santiago de Compostela (Spain) and a member of the Novos Medios research group. His research interests include risk communication, misinformation and the application of computational methods, big data, and artificial intelligence to the study of health information.

Mónica Pérez Ríos is Professor of Epidemiology and Public Health at the University of Santiago de Compostela (Spain). Department of Preventive Medicine and Public Health. Consortium for Biomedical Research in Epidemiology and Public Health (CIBER en Epidemiología y Salud Pública/ CIBERESP). Health Research Institute of Santiago de Compostela (Instituto de Investigación Sanitaria de Santiago de Compostela—IDIS).

Tanja Perko has a PhD in risk communication and risk perception from the University of Antwerp (Belgium), following her undergraduate studies in Journalism and Defense Studies (B.A.) and a Master's degree in Political Sciences (MSc) from the University of Ljubljana, Slovenia. She is currently a Senior Researcher at SCK CEN and University Antwerp, a promotor and mentor for PhD students and a guest lecturer at various universities and international organisations.

Alberto Ruano Ravina is Professor of Epidemiology and Public Health at the University of Santiago de Compostela (Spain). He is affiliated to the Department of Preventive Medicine and Public Health. Consortium for Biomedical Research in Epidemiology and Public Health (CIBER en Epidemiología y Salud Pública/CIBERESP), the Health Research Institute of Santiago de Compostela (Instituto de Investigación Sanitaria de Santiago de Compostela—IDIS) and the Cross-disciplinary Research Center in Environmental Technologies (CRETUS).

Caroline Schieber is project leader at CEPN (Nuclear Protection Evaluation Center, France). Her work mainly deals with the practical implementation of the principle of radiation protection optimisation, the development of radiation protection culture and the management of radon at work and in dwellings.

Yevgeniya Tomkiv is a researcher at Norwegian University of Life Sciences (NMBU), Norway. Her research interests encompass risk communication about ionising radiation, societal aspects of radiological emergencies, radon and naturally occurring radioactive material and stakeholder involvement in environmental remediation and natural resource management.

Carlos Toural-Bran has a degree in Journalism (2006) and a PhD in Communication Sciences (2013). He is a full professor at the Department of Communication Sciences at the Universidade de Santiago de Compostela (Spain). In the field of management, he held the position of Vice-Dean of the Faculty of Communication Sciences on two occasions, as well as secretary of the Novos Medios Research Group. He currently heads the USC Chancellor's Office.

Leonor Varela Lema is Associate Professor of Epidemiology and Public Health at the University of Santiago de Compostela (Spain). She is affiliated to the Department of Preventive Medicine and Public Health, the Consortium for Biomedical Research in Epidemiology and Public Health (CIBER en Epidemiología y Salud Pública/CIBERESP) and the Health Research Institute of Santiago de Compostela (Instituto de Investigación Sanitaria de Santiago de Compostela—IDIS).

Jorge Vázquez-Herrero, PhD in Communication, Universidade de Santiago de Compostela (Spain). He is currently Professor at the Department of Communication Sciences and member of Novos Medios research group (USC). He was Visiting Scholar at Rosario, Leeds, Tampere and Utrecht. His research focuses on the impact of technology and platforms in digital journalism and narratives.

Ángel Vizoso has a PhD in Communication, is Assistant Professor at the Department of Communication Sciences and a researcher at Novos Medios Research Group of the Universidade de Santiago de Compostela (Spain). He is editor at Revista de la Asociación Española de Investigación de la Comunicación (RAE-IC). His main lines of research are focused on the study of the visualisation and verification of information and changes in professional profiles.

Jessica Zorogastua Camacho is Assistant Professor at the Universidad Rey Juan Carlos (Spain) and research director of the Luca de Tena Foundation. She has more than 20 years of professional experience in communication consultancy to senior officials of the Spanish Public Administration and communication management of private companies and foundations.

Foreword

Risk communication is a core component of public health that requires planning, preparation and practice to be effective. It involves an interactive process in which experts and Government officials exchange information with the public, not merely to raise awareness of looming threats but to foster comprehension and encourage protective measures.

When it comes to a carcinogenic indoor air contaminants such as radon, a radioactive gas that infiltrates our homes and workplaces, risk communication is complicated by both perception and emotion. Despite the World Health Organization's recognition of radon as the second cause of lung cancer right after smoking and as the primary cause in non-smokers, most people do not perceive it as a significant threat, as it is odourless, invisible and of natural origin. Countries with many-year-running national programmes against radon often report difficulties in engaging the public to measure it or to remediate their homes when elevated radon levels are found, even if incentives are offered.

However, when the radon issue grabs headlines for the first time in a country, it unleashes outrage and anxiety. Fortunately, with the right communication strategy, this turmoil can be transformed into effective action, laying the groundwork for a national strategy against radon. The battle against radon is one of sustained endurance. Once the initial interest has waned and radon has retreated from the spotlight but awareness and access to measurement and mitigation services have increased, effective communication becomes crucial in sustaining motivation to take action.

Since the International Agency on Cancer Research recognized radon as a Group I carcinogen in 1988, we have come a long way: we now possess a much better knowledge of the health effects of radon exposure; access to high-quality radon measurement services is widespread in many countries; a wealth of international expertise has been accumulated in radon mitigation; and an increasing number of national building codes include provisions for radon-resistant new homes. Yet progress in radon risk communication has not kept pace with developments in other areas.

Initially, radon risk communication fell within the purview of radiation authorities. They assessed the scope of the problem by creating radon maps and aided in bolstering national measurement and mitigation capabilities, but their often cautious approach to public communication aimed at avoiding alarm and hindered widespread dissemination.

Throughout the 2000s, the efforts of the WHO, the IAEA and other international players successfully emphasized the need to tackle the radon problem in an integrated and collaborative manner. They urged countries to develop coordinated national strategies in which communication plays a fundamental role. Many Governments heeded this call and articulated national radon plans, with communication taking centre stage. But in our ever-evolving digital landscape, communication strategies, once more, might fall behind.

The 21st century has ushered in a dramatic transformation in how citizens acquire information and shape their opinions, primarily driven by the emergence of the internet and, in particular, social networks.

Today, we are flooded with an excessive amount of information, often causing critical messages to go unnoticed. Digital communication predominantly occurs within private spaces and is disseminated behind the public scene. This environment fosters the proliferation of misinformation and erodes trust in both scientific authority and public sources. WHO experts advocate for improving health-related content in mass media and increasing digital and health literacy as a means to combat the infodemic.

This book is a valuable step to promoting effective radon communication in today's interconnected world. It brings together the collective expertise and insights of a diverse group of professionals from different fields, all with the objective of providing a comprehensive international perspective on radon-related initiatives and success stories, along with an analysis of the current presence of radon-related content across mainstream and social media.

Furthermore, the book contributes to a better understanding of the challenges associated with radon communication in the digital age and advocates for public authorities to enlist mass media support and actively engage with the public on social media platforms. In this new paradigm, citizens are viewed as vital partners capable of understanding scientific information and collaborating towards the shared goal of reducing radon-induced lung cancers.

Marta García-Talavera
Head of the Natural Radiation Section
Nuclear Safety Council (Spain)

Acknowledgements

This book has been prepared as part of the activities developed in:

- The research project *Radon in Spain: Public perception, media agenda and risk communication* (RAPAC), financed by the Spanish Nuclear Safety Council [Consejo de Seguridad Nuclear] (SUBV-13/2021).
- Novos Medios research group, Universidade de Santiago de Compostela (Spain).

1 An introduction to radon gas and risk communication

José Sixto-García, Sara Pérez-Seijo and Berta García-Orosa

Public health and risk communication

The primary objectives of public health encompass safeguarding the population's well-being through the promotion of healthy lifestyles and the advocacy of disease prevention and health protection programs (Schober et al., 2022; Meghani, 2022; Marchell et al., 2022; Myhre et al., 2022). However, for public health programs to achieve success, it is imperative that public institutions (Paniagua, 2022) effectively communicate the inherent risks and the necessity for citizens to actively adopt precautionary measures and cooperation.

Risk communication is an emerging field in communication studies that addresses the need to investigate the type of information that should be conveyed to the public during crises, disasters, or hazards (Farré, 2005). The inception of risk communication dates back to the late 1970s in the United States, particularly when nuclear and chemical industries sought to mitigate the sense of fear and uncertainty their activities generated among the populace. However, the current conceptualization of this sort of communication focuses on a more dialogic and interactive engagement with the public. This approach aims to empower individuals exposed to risks to make informed decisions that mitigate the threat or risk and take necessary preventive measures (Idoiaga et al., 2016; Valente et al., 2021). Public risk communication cannot be comprehended or executed without ensuring that individuals understand and appreciate the risks to protect themselves and their surroundings (Johansson et al., 2021). Therefore, it is crucial to develop and implement effective, culturally appropriate health messages (Boyd & Furgal, 2022).

For many years, research has been carried out on the factors that influence whether a phenomenon is perceived as a risk or not. Over 30 years ago, an empirical study investigated the functional relationships among five sets of variables involved in the amplification process (Renn et al., 1992): the amount of press coverage, physical consequences, public responses, individual layperson perceptions, and the socioeconomic and political impacts. Renn et al. (1992) discovered that perceptions and social responses are more strongly correlated with

DOI: 10.4324/9781032618180-1

risk exposure than with its magnitude. Thus, only multidisciplinary research will be capable of addressing the phenomenon comprehensively, taking into account not only its technical dimensions but also the public's perception—the sole element that implicates action, prevention, and mitigation.

Similarly, scientific literature offers numerous studies on risk communication in crisis situations (Boyd & Furgal, 2019; Gerdes, 2022). However, it posits that to gain a comprehensive understanding of risk in the context of a crisis, it is essential to analyze a risk that persists over time, such as radon gas—the fundamental subject of study in the chapters presented in this book.

The case of radon gas

However, what occurs in the specific case of radon gas? Is it being communicated that it poses a problem for public health? Despite radon being a natural gas formed through the decay of uranium in the soil, scientific evidence demonstrates that it poses a public health risk as it ranks as the second leading cause of lung cancer worldwide, following smoking (Bouder et al., 2021). It is the primary cause in non-smokers (García-Talavera & López, 2019; Neri et al., 2018). The proportion of lung cancer cases associated with radon, relative to the total, is estimated to vary between 3% and 14%, depending on the country's average radon concentration and the calculation method employed.

The primary source of the population's exposure to ionizing radiation is natural radon and its decay products. As a tasteless, odourless, invisible, and naturally occurring gas, radon is undetectable without testing. Buildings constructed in areas with bedrock or soils rich in uranium can develop high concentrations of radon in indoor air, posing a significant health threat (Ryan et al., 2015).

Since 1998, the World Health Organization (WHO) has classified radon as a carcinogenic element when there is exposure to high levels indoors (García-Talavera et al., 2013). Simultaneously, it emphasizes the importance of public awareness regarding proven and durable methods to prevent leaks in new construction and to reduce radon concentrations in existing homes (WHO, 2021). Its significance lies in being a sustained risk over time, with severe health effects demonstrated for almost 40 years. Moreover, there is limited public awareness of its gravity despite the available knowledge (Khan & Chreim, 2019).

From a historical perspective, the study of radon risk exposure took its initial steps through the coordinated efforts of the International Commission on Radiological Protection (ICRP) in the late 1950s among uranium miner communities and has evolved since then. Between 1928 and 1934, the ICRP formulated some recommendations regarding whole-body and limb exposure to radon gas (Lopes et al., 2021).

Health-related environmental issues associated with radon accumulation in factories in Iraq (Othman et al., 2022), the presence of radon indoors and its diurnal variation in Cameroon (Sadjo et al., 2022), the accumulation of radon in drinking water in Pakistan (Shakoor et al., 2022), or radon levels in the United States (Carrion-Matta et al., 2021) are some of the matters recently investigated by the scientific community. These studies underscore the global significance of the public health problem posed by radon.

Given this context, both the World Health Organization and the European Union have been emphasizing the importance of communication about this public health risk for several years, particularly in areas with higher incidence levels.

Communicating radon gas: strategies, initiatives, and public opinion

Awareness of the dangers posed by radon cannot be achieved without understanding the risks of exposure to it (Khan et al., 2019). Therefore, the main objective of this book, titled *Communicating Public Health: Risk Communication Strategies and Public Perceptions of Radon Gas*, is to analyze the risk communication strategies currently employed to communicate public health and its risks, with a specific focus on the case of radon gas. This work is conceived in a context where societal knowledge of radon risks is limited, and the presence of the topic in traditional and digital native media is low and ineffective (Bouder et al., 2021).

Now is the opportune moment to publish a book like this, as there is a societal demand to learn more about radon, as demonstrated by recent studies conducted in Europe (Makedonska et al., 2018; Lofsted, 2019; Cori et al., 2022). These highlight the necessity of addressing an unresolved problem, a need that had already been identified by earlier research conducted in the United States (Fisher et al., 1991; Page, 1994). On the other hand, recent works consistently emphasize risk communication as a key aspect of the public health response (Thomas et al., 2022). Though health risks often make front-page news (Renn, 2006), communication efforts regarding radon risks continue to be ineffective (Bouder et al., 2021).

The volume *Communicating Public Health: Risk Communication Strategies and Public Perceptions of Radon Gas* is approached from a cross-cutting and multidisciplinary perspective, incorporating contributions from 33 authors representing nine countries—Spain, Belgium, France, Norway, Ireland, Hungary, Portugal, Argentina, and Ecuador—and diverse areas of expertise, including communication, journalism, physics, and citizen science. Specifically, the book is divided into three distinct parts.

In the first part, it is explained why radon gas constitutes a public health problem, details its consequences on people's lives, discusses associated risks, and emphasizes the relevance of citizen science. On the one hand, Lucía Martín de Bernardo Gisbert, Mónica Pérez Ríos, Leonor Varela Lema, Juan Miguel Barros Dios, and Alberto Ruano Ravina present research stemming from the first author's doctoral thesis. Titled *Indoor radon as a public health problem: available evidence on radon and its health effects*, this chapter deals with the public health issue of radon in enclosed spaces, such as homes or workplaces. The scholars highlight the fact that radon causes cancer and the need to reduce exposure both in the general population and among workers. On the other hand, Meritxell Martell, Tanja Perko, Sylvain Andresz, Caroline Schieber, Yevgeniya Tomkiv, Alison Dowdall, Leo McKittrick, and Veronika Oláhné Groma examine radon gas from the perspective of citizen science—an approach seeking to understand how the public collaborates with radon experts during the research process. In this chapter, the authors explore the potential of citizen science in preventing and mitigating the effects of radon gas. Through various pilot projects conducted in four different countries, they conclude that citizen science will be crucial in reducing the risk of lung cancer, the primary disease caused by radon gas.

In the second part, the relevance of communicating public health in digital media is discussed, in particular by utilizing data visualization tools and disruptive technologies such as virtual reality to engage with citizens and adapt to the logic and consumption patterns of diverse audiences. First, Carmen Costa-Sánchez, Sofía Gomes, and Xosé López-García review prominent journalistic initiatives in the field of public health specialization in Latin America. These digital media outlets focus on the general public and/or healthcare professionals, aiming to connect with their publics through various digital platforms in a scenario marked by growing misinformation. Second, Ángel Vizoso, Gabriela Coronel-Salas, and Carlos Toural-Bran assess the importance of data visualization and information to ensure the effectiveness of risk communication. The authors start from the premise that information related to risk communication can be complex and challenging to comprehend for non-specialized audiences. Therefore, from a journalistic perspective, media outlets should opt for visual genres and formats that are effective in alerting and raising awareness among citizens. Third, Pavel Sidorenko Bautista, Jessica Zorogastua Camacho, and Mariola Moreno Calvo evaluate the potential of virtual reality experiences and the metaverse for public administrations to design information and awareness campaigns regarding radon gas and its impact on public health. The scholars contend that immersive media can serve as appealing alternatives to reach certain segments of the population, especially the youngest, aiming to provide them with useful and relevant information to limit or minimize the effects of the risk.

Finally, in the third part, the public perception of radon risks is explored based on media coverage and expert opinions in the field. María-Cruz Negreira-Rey, Jorge Vázquez-Herrero, and Rita Araújo open this section with a chapter dedicated to the inclusion of radon gas in the media agenda of local journalistic outlets in Spain. The authors conclude that the local nature of the news can play a key role in risk perception because information becomes more significant for citizens when it informs them about regulatory changes that directly affect them or the impact of gas in their territory or the areas they frequent. Following this chapter, Lila Luchessi and Pablo Escandón-Montenegro specifically focus on public perception and the treatment of risk communication in Latin America. These researchers examine and compare how the media communicates risk in Ecuador and Argentina, estimating that, in general, crises and their associated dangers are more often linked to human actions than to consequences for the territory and the environment. Lastly, Noel Pascual-Presa and Tania Forja-Pena conclude the book by addressing radon gas from a communicative perspective as a fundamental axis for the prevention and mitigation of public health risks. Their work centres on the analysis of two main actors: the media and expert information sources. In addition to a review of international scientific literature, the chapter studies the coverage of radon gas in news outlets and also analyzes the opinion of the scientific sector regarding the dissemination and communication of the problem that radon gas represents for public health.

Acknowledgments

This chapter is part of the project *Radon in Spain: Public perception, media agenda and risk communication* (RAPAC), financed by the Spanish Nuclear Safety Council [Consejo de Seguridad Nuclear] (SUBV-13/2021).

References

Bouder, F., Perko, T., Lofstedt, R., Renn, O., Rossmann, C., Hevey, D., Siegrist, M., Ringer, W., Pölzl-Viol, C., Dowdall, A., Fojtíková, I., Barazza, F., Hoffmann, B., Lutz, A., Hurst, S., & Reifenhäuser, C. (2021). The Potsdam radon communication manifesto. *Journal of Risk Research, 24*(7), 909–912. https://doi.org/10.1080/13669877.2019.1691858

Boyd, A. D., & Furgal, C. M. (2019). Communicating environmental health risks with indigenous populations: A systematic literature review of current research and recommendations for future studies. *Health Communication, 34*(13), 1564–1574. https://doi.org/10.1080/10410236.2018.1507658

Boyd, A. D., & Furgal, C.M. (2022). Towards a participatory approach to risk communication: The case of contaminants and Inuit health. *Journal of Risk Research, 25*(7), 892–910. https://doi.org/10.1080/13669877.2022.2061035

Carrion-Matta, A., Lawrence, J., Kang, C., Wolfson, J., Longxiang, L., Zilli, C., Schwartz, J., Demokritou, P., & Koutrakis, P. (2021). Predictors of indoor radon levels in the Midwest United States. *Journal of the Air & Waste Management Association, 71*(12), 1515–1528. https://doi.org/10.1080/10962247.2021.1950074

Cori, L., Bustaffa, E., Cappai, M., Curzio, O., Dettori, I., Loi, N., Nuchis, P., Sanna, A., Serra, G., Sirigu, E., Tidore, M., y Bianchi, F. (2022). The role of risk communication in radon map-ping, risk assessment and mitigation activities in Sardinia (Italy). *Advances in Geosciences, 57*, 49–61. https://doi.org/10.5194/adgeo-57-49-2022

Farré, J. (2005). Comunicación de riesgo y espirales del miedo. *Comunicación y Sociedad, 3*, 95–119.

Fisher, A., McClelland, G. H., Schulze, W. D., & Doyle, J. K. (1991). Communicating the risk from radon. *Journal of the Air & Waste Management Association, 41*(11), 1440–1445. https://doi.org/10.1080/10473289.1991.1046694

García-Talavera, M., & López, F. J. (2019). Cartografía del potencial de radón en España. *Consejo de Seguridad Nuclear.* https://bit.ly/3UXMvRV

García-Talavera, M., Martín, J. L., Gil, R., García, J. P., & Suárez, E. (2013). El mapa predictivo de exposición al radón en España. *Consejo de Seguridad Nuclear.* https://bit.ly/3VhVKMf

Gerdes, J. (2022). Diagnosing unsettled stasis in transnational communication design: An exploration of public health emergency communication. *Technical Communication Quarterly,* 1–16. https://doi.org/10.1080/10572252.2022.2069286

Idoiaga, N., Gil De Montes, L., & Valencia, J. F. (2016). Communication and representation of risk in health crises: The influence of framing and group identity. *International Journal of Social Psychology, 31*(1), 59–74. https://doi.org/10.1080/02134748.2015.1101313

Johansson, B., Lane, D., Sellnow, D., & Sellnow, T. (2021). No heat, no electricity, no water, oh no!: An IDEA model experiment in instructional risk communication. *Journal of Risk Research, 24*(12), 1576–1588. https://doi.org/10.1080/13669877.2021.1894468

Khan, S. M., & Chreim, S. (2019). Residents' perceptions of radon health risks: A qualitative study. *BMC Public Health, 19*(1), 1–11.

Khan, S. M., Krewski, D., Gomes, J., & Deonandan, R. (2019). Radon, an invisible killer in Canadian homes: Perceptions of Ottawa-Gatineau residents. *Canadian Journal of Public Health, 110*(2), 139–148. https://doi.org/10.17269/s41997-018-0151-5

Lofstedt, R. (2019). The communication of radon risk in Sweden: Where are we and where are we going? *Journal of Risk Research, 22*(6), 773–781. https://doi.org/10.1080/13669877.2018.1473467

Lopes, S. I., Nunes, L. J., & Curado, A. (2021). Designing an Indoor Radon Risk Exposure Indicator (IRREI): An evaluation tool for risk management and communication in the IoT age. *International Journal of Environmental Research and Public Health, 18*(15), 7907. https://doi.org/10.3390/ijerph18157907

Makedonska, G., Djounova, J., & Ivanova, K. (2018). Radon risk communication in Bulgaria. *Radiation Protection Dosimetry, 181*(1), 26–29. https://doi.org/10.1093/rpd/ncy096

Marchell, T. C., Santacrose, L., Laurita, A., & Allan, E. (2022). A public health approach to preventing hazing on a university campus. *Journal of American College Health,* online first. https://doi.org/10.1080/07448481.2021.2024210

Meghani, Z. (2022). The impact of vertical public health initiatives on gendered familial care work: Public health and ethical issues. *Critical Public Health, 32*(5), 600–607. https://doi.org/10.1080/09581596.2021.1908960

Myhre, S. L., French, S. D., & Bergh, A. (2022). National public health institutes: A scoping review. *Global Public Health, 17*(6), 1055–1072. https://doi.org/10.1080/17441692.2021.1910966

Neri, A., McNaughton, C., Momin, B., Puckett, M., y Gallaway, M. S. (2018). Measuring public knowledge, attitudes, and behaviors related to radon to inform cancer control activities and practices. *Indoor Air, 28*(4), 604–610. https://doi.org/10.1111/ina.12468

Othman, S., Ahmed, A., & Mohammed, S. (2022). Environmental health risks of radon exposure inside selected building factories in Erbil city, Iraq. *International Journal of Environmental Analytical Chemistry* (online first). https://doi.org/10.1080/03067319.2022.2107923

Page, S. D. (1994). Indoor radon: A case study in risk communication. *American Journal of Pre-ventive Medicine, 10*(3), 15–18. https://doi.org/10.1016/S0749-3797(18)30545-2

Paniagua, P. (2022). Elinor Ostrom and public health. *Economy and Society, 51*(2), 211–234. https://doi.org/10.1080/03085147.2022.2028973

Renn, O. (2006). Risk communication—Consumers between information and irritation. *Journal of Risk Research*, 9(8), 833–849. https://doi.org/10.1080/13669870601010938

Renn, O., Burns, W. J., Kasperson, J. X., Kasperson, R. E., & Slovic, P. (1992). The social amplification of risk: Theoretical foundations and empirical applications. *Journal of Social Issues*, 48(4), 137–160. https://doi.org/10.1111/j.1540-4560.1992.tb01949.x

Ryan, P., Muller, N., & Munroe, D. (2015). Radon risk and public health in Vermont. *Middlebury College Environmental Studies Senior Seminar Spring 2015*. https://neaarst.org/wp-content/uploads/2015/07/VermontRadonPolicyResearch-PaperGilman-Roe.pdf

Sadjo, T., Hamadou, Y., Valentin, S., & Mohamadou, A. (2022). Soil gas radon, indoor radon and its diurnal variation in the northern region of Cameroon. *Isotopes in Environmental and Health Studies*, online first. https://doi.org/10.1080/10256016.2022.2102617

Schober, D. J., Carlberg-Racich, S., & Dirkes, J. (2022). Developing the public health workforce through community-based fieldwork. *Journal of Prevention & Intervention in the Community*, 50(1), 1–7. https://doi.org/10.1080/10852352.2021.1915736

Shakoor, H., Jehan, N., Khan, S., & Khattak, N. (2022). Investigation of Radon Sources, Health Hazard and Risks assessment for children using analytical and geospatial techniques in District Bannu (Pakistan). *International Journal of Radiation Biology*, 98(6), 1176–1184. https://doi.org/10.1080/09553002.2021.1872817

Thomas, M., Kaufman, S., Klemm, C., & Hutchins, B. (2022). The co-evolution of government risk communication practice and context for environmental health emergencies. *Journal of Risk Research* (online first). https://doi.org/10.1080/13669877.2022.2077414

Valente J. -P., Gouveia, C., Neves, M. -C., Vasques, T., & Bernardo, F. (2021). Small town, big risks: Natural, cultural and social risk perception. *PsyEcology*, 12(1), 76–98. https://doi.org/10.1080/21711976.2020.1853946

WHO. (2021). *Radon and health*. www.who.int/news-room/fact-sheets/detail/radon-and-health

Part I

Radon

A public health issue

2 Indoor radon as a public health problem. Available evidence on radon and its health effects

Lucía Martín de Bernardo Gisbert[1],
Mónica Pérez Ríos, Leonor Varela Lema,
Juan Miguel Barros Dios
and Alberto Ruano Ravina

What is indoor radon exposure?

Radon exposure refers to the exposure via inhalation to radon gas (^{222}Rn) and its decay products (namely ^{218}Po and ^{214}Po), which are often found as small solid particles in suspension (IARC, 1988). Radon and its decay products are part of the Uranium-238 (^{238}U) radioactive decay chain; thus, they ultimately originate from the uranium present in the earth's crust.

Radon gas is naturally exhaled from the bedrock into the atmosphere. There, it dilutes and remains at relatively low concentrations. Alternatively, radon can enter indoors, inside buildings or underground mines and accumulate, reaching high concentrations that pose a risk to health (World Health Organization, 2009). Therefore, from a public health perspective, indoor radon exposure refers to human exposure to radon gas and its decay products within enclosed spaces.

Reference levels of radon

Radon gas is naturally exhaled from the bedrock into the atmosphere. There, it dilutes and remains. A reference level or action level of radon is the concentration of radon that shall not be exceeded in homes, workplaces, education facilities and any other indoor space with public permanence. Many countries and supranational institutions have established reference levels of radon concentration, ranging from 100 to 300 Bq/m3.

The first reference level was provided in 1987 by the US Environmental Protection Agency (EPA), which established an action level of 148 Bq/m3 (Environmental Protection Agency, 1987). Since 2009, the World Health Organization (WHO) recommends countries implement an action level below 300 Bq/m3, ideally of 100 Bq/m3(World Health Organization, 2009). Finally, the European Union (UE), in Directive 13/59/Euratom, established a reference level equal to or below 300 Bq/m3 of radon concentration (Council Directive, 2013/59/

DOI: 10.4324/9781032618180-3

Euratom of 5 December 2013 Laying down Basic Safety Standards for Protection against the Dangers Arising from Exposure to Ionising Radiation, 2013).

While many EU countries adopted the maximum reference level allowed by the Directive (300 Bq/m3), some others have established reference levels below that; this is the case of Ireland, with a reference level of 200 Bq/m3 (*Radon Map | Environmental Protection Agency, n.d.*). Other international societies, such as the International Commission of Radiological Protection (ICRP), recommend a reference level of 300 Bq/m3 (ICRP, 2014). Of note, many countries or organizations have reduced the reference level as more scientific evidence was available on radon and lung cancer. This is the case of the ICRP, Canada, Spain and other countries and institutions (Ruano-Ravina, Kelsey, et al., 2017). On the other hand, many countries do not have a national radon reference level at all. For instance, a review in 2020 confirmed nine countries in Central and South America did not have one (Giraldo-Osorio et al., 2020).

Ionizing radiation in the lungs

Radon and its decay products are the main source of ionizing radiation in the general population in many countries (Radon Causes Most of the Radiation Received by Finns | Säteilyturvakeskus (STUK), n.d.; UK radon, n.d.). This is because every human is exposed to radon, commonly at very low levels, as it is a natural part of the air we breathe. Radon and its decay products emit ionizing radiation in the form of alpha particles (alpha radiation) as part of the decay process. As shown in Table 2.1, alpha radiation occurs when Radon-222 decays into Polonium-218 (half of the ^{222}Rn particles will do so within 3.8 days), when Polonium-218 decays into Lead-214 (half will do so within 3 min), or when Polonium-214 decays into Lead-210 (half will do so within 0.0002 seconds).

Alpha particles have low penetration and are highly ionizing. This means they cannot pass through tissues, only producing ionization at the closest spot they encounter. When radon and its decay products enter the respiratory tract, some will emit alpha particles that impact the pulmonary cell lining and could produce mutations when they hit the cell. These mutations could eventually lead to tumour formation if not repaired by different cellular mechanisms.

Table 2.1 Extract of Radon-222 decay series.

Decay product	Half-life	Radiation type	Decay direction
Radon-222	**3.8 days**	**alpha**	↓
Polonium-218	**3 mins**	**alpha**	↓
Lead-214	29 mins	beta	↓
Bismuth-214	18 mins	beta	↓
Polonium-214	**0,0002 secs**	**alpha**	↓
Lead-210	22 years	beta	↓

Source: (IARC, 1988).

Of note, approximately half of the ionizing radiation received by human beings comes from indoor radon exposure (Ruano-Ravina & Wakeford, 2020). Nevertheless, this percentage may vary depending on the amount of ionizing radiation coming from artificial sources, mainly medical imaging testing. In the USA, for example, the radiation received from artificial sources has surpassed that received from natural sources. There is a public health concern regarding individuals with high radon concentrations at home who also undergo frequent ionizing radiation-based imaging tests, as this combination may lead to exceptionally high overall radiation exposure (Ruano-Ravina & Wakeford, 2020).

Cohorts of underground miners: BEIR VI report

In 1988, radon and its decay products were classified as a group 1 human carcinogen by the International Agency for Research on Cancer (IARC) (IARC, 1988). IARC Monograph on Radon concluded there was sufficient evidence available in humans at that time. The epidemiological evidence arose mainly from cohort studies of underground miners, especially uranium miners. These cohorts were followed up from the 1950s, and by the 80s, they had provided definitive or interim results on the increased lung cancer risk associated with radon exposure (IARC, 1988).

After 1988, numerous underground miner studies (new ones or already existing ones) continued providing further evidence that was pooled together in the US National Research Council's report of the Sixth Committee on Biological Effects of Ionizing Radiations (BEIR VI). BEIR VI report was published in 1999 and provides a comprehensive risk model for the relationship between radon exposure and lung cancer risk for indoor radon exposure that is still in use today. The model was meant for the general population but inferred from results of 11 cohorts of underground miners: eight cohorts of uranium miners from West Bohemia (Czech Republic), Ontario, Beaverlodge and Port Radium (Canada), Colorado Plateau and New Mexico (USA), France and Radium Hill (Australia); plus, one cohort of iron miners in Malmberget (Sweden), one of tin miners in Yunnan (China) and one of fluorspar miners in Newfoundland (Canada) (National Research Council, 1999).

BEIR VI report assumes that radon exposure and lung cancer risk follow a linear dose-response with no threshold. They propose two equally valid models: the exposure-age-concentration model (EAC model) and the exposure-age-duration model (EAD model). They simply differ on whether they use average radon concentration or radon exposure duration. These two models are still widely used to estimate radon-attributable mortality, especially in Canada and the USA (Martin-Gisbert et al., 2022).

The exposure to radon in the general population, mainly residential exposure, is several orders of magnitude below underground miners' exposure. However, the least exposed miners and most exposed population can have equivalent radon exposure. BEIR VI risk models assumed a linear-non-threshold (LNT) relationship between radon exposure and lung cancer risk. As a result of assuming the LNT hypothesis, the same linear dose-response observed in high exposures can be expected at low exposures. Extrapolating a miner-based risk model to the general population has numerous limitations: first, an exposure threshold could not be discarded; and secondly, the characteristics of cohorts of miners and the data available raised further concerns. The cohort population was only composed of males of working age; thus, no data was available on the possible effects on children and women. Also, miners were exposed to other carcinogens, most importantly arsenic and tobacco, and both were poorly reported in the cohort studies; only six out of the 11 cohorts included some data on smoking habit. Finally, other different dusts, high physical activity and a varied range of protection measures and ventilations used made the extrapolation of results to the general population difficult.

At that moment, evidence from studies carried out in the general population was scarce and inconclusive. Eight case-control studies were available at that time; however, these were insufficient to provide a risk model. Furthermore, the BEIR VI report explicitly mentioned that such evidence might be impossible provided a low relative risk expected (RR = 1.1 per each 148 Bq/m3 increase), a skewed radon distribution and uncertainties related to residence mobility. Thus, in order to detect a dose-response relationship between residential radon exposure and lung cancer, evidence should arise from areas with high radon exposure, low residence mobility and numerous cases and controls, probably by pooling several case-control studies (National Research Council, 1999).

Evidence in the general population

In 2001, a review on radon and lung cancer was published with a very declarative title *Radon: A likely a carcinogen at all exposures* (Darby et al., 2001). This meant that low exposures to radon occurring across the general population could be carcinogenic. However, the word "likely" in this title emphasizes the limitations of the existing evidence to confirm this. These limitations, detailed in the previous paragraph, are derived from extrapolating miners' risk models to the general population, plus the insufficient data from case-control studies in the general population at that time.

In order to elucidate whether residential indoor radon was a risk factor for lung cancer in the general population, several case-control studies were carried out in different countries (Auvinen et al., 1996; JM. Barros-Dios et al., 2002; Field et al., 2000; Létourneau et al., 1994) and most found an association between residential radon and lung cancer. Most of them measured radon with passive detectors in homes with cases of lung cancer and controls, using long-term measurements. Remarkably, one case-control study found an association between indoor radon and lung cancer at very low levels of exposure, as low as 37 Bq/m3, thus supporting the LNT hypothesis (Barros-Dios et al., 2002).

As mentioned in the BEIR VI report, only pool studies could have enough sampling to detect the increased risk of lung cancer due to residential radon exposure. This need was addressed with the publication of the results from three different pooling studies: one Chinese (Lubin et al., 2004), one North American (Krewski et al., 2005) and one European (Darby et al., 2005, 2006).

The Chinese pooling (Lubin et al., 2004) analyzed data from two large case-control studies, one from the city of Shenyang and one from Gansu province. The pool comprised 1,050 lung cancer cases and 1,996 controls. The odds ratio (OR) for lung cancer of those exposed to 100 Bq/m3 of radon concentration at home was 1.13, with a 95% confidence interval (95% CI) ranging from 1.01 to 1.36.

The North American pooling (Krewski et al., 2005) analyzed data from seven case-control studies in Winnipeg (Canada) and in Connecticut, Iowa, Missouri, New Jersey and Utah-South Idaho (USA), comprising a total of 3,662 cases and 4,966 controls. The OR for lung cancer of those exposed to 100 Bq/m3 of radon concentration at home was 1.11 (95% CI 1.00–1.28).

Finally, the European pooling (Darby et al., 2005, 2006) analyzed data from 13 case-control studies from Austria, Czech Republic, Finland, France, Germany, Italy, Spain, Sweden and the UK. This is the largest pool, comprising a total of 7,148 cases and 14,208 controls across Europe. The Relative Risk (RR) reported at 100 Bq/m3 was 1.16 (95% CI 1.05–1.31). Furthermore, this European pooling study is the most cited study on radon and lung cancer.

It is important to highlight that in these pooling studies, all principal investigators from each case-control study included sent the entire database with all participants (cases and controls) to a coordinating centre. For instance, Oxford University, under the guidance of Prof. Sarah Darby, served as the coordinating centre for the European pooling study. This collaborative effort allowed for in-depth research involving thousands of participants.

These three pooling studies yielded equivalent results that were compatible with a linear non-threshold relationship between residential radon exposure and lung cancer risk. More specifically, for each 100 Bq/m3 increase in residential indoor radon concentration, the lung cancer risk increased 13%, 11% and 16% in the Chinese, North American and European pooling respectively. Furthermore, these results were completely coherent with the extrapolations from miner-based risk models in BEIR VI, thus confirming residential indoor radon exposure is a risk factor for lung cancer.

WHO Handbook on Indoor Radon

As a result of the conclusive evidence available after the publication of the pooling studies, the World Health Organization launched the International Radon Project in 2005, with the participation of experts and policymakers from more than 30 countries. The objective of the project was to bring a global public health perspective and to provide sound policy strategies to protect the population from the risks arising from radon exposure. The main deliverable of the project was the *WHO Handbook on Indoor Radon: A public health perspective*, published in 2009 (World Health Organization, 2009).

The WHO handbook from 2009 is still the main reference on the topic today. First, it includes a comprehensive summary and analysis of the evidence available on radon health effects and provides key conclusions, remarkably concluding that radon is the main cause of lung cancer after smoking in many countries. Furthermore, it emphasizes that there is no known threshold of radon concentration free of risk, thus, there is no safe level of radon.

Secondly, it provides a common framework for the countries on how to actually enact effective strategies to protect the population from radon health risks: how to measure indoor radon exposure, how to build new radon-free buildings, how to mitigate radon in existing ones, how to communicate radon risks to the population and how to establish a national reference level of radon concentration, regulations or a national radon map. This common framework was the basis for many public policies and regulations, such as the Directive 2013/59/Euratom in the EU (Council Directive 2013/59/Euratom of 5 December 2013 Laying down Basic Safety Standards for Protection against the Dangers Arising from Exposure to Ionising Radiation, 2013).

Radon and tobacco: a deadly couple

It provides a common framework for the countries on how to actually enact effective strategies. Tobacco consumption is the main cause of lung cancer, responsible for 80 to 90% of lung cancer incidence (Hansen et al., 2020; Parkin, 2011). The European pooling study observed that radon exposure increased lung cancer risk equally among smokers and never smokers; each 100 Bq/m3 increase in residential radon levels increases 16% their risk of lung cancer (Darby et al., 2006). However, the baseline risk of lung cancer is extremely different in smokers and non-smokers.

Tobacco use and radon exposure interact, posing very high risks for radon-exposed smokers. For instance, in the European pooling, a concentration of 400 Bq/m3 at home increased lung cancer risk 42 times in smokers and 1.6 in never smokers (see Figure 2.1) (Darby et al., 2006). This synergic interaction between radon exposure and tobacco use has also been observed in further case-control studies (Barros-Dios et al., 2012; Lorenzo-Gonzalez et al., 2020).

Currently, most radon-induced lung cancers occur among smokers (World Health Organization, 2009). For this reason, *WHO Handbook on Indoor Radon* encourages including tobacco in the radon prevention agenda.

Finally, secondhand tobacco smoke has been a declared carcinogen that increases lung cancer risk since 2004 (IARC Working Group on the Evaluation of Carcinogenic Risks to Humans et al., 2004). There are no studies available

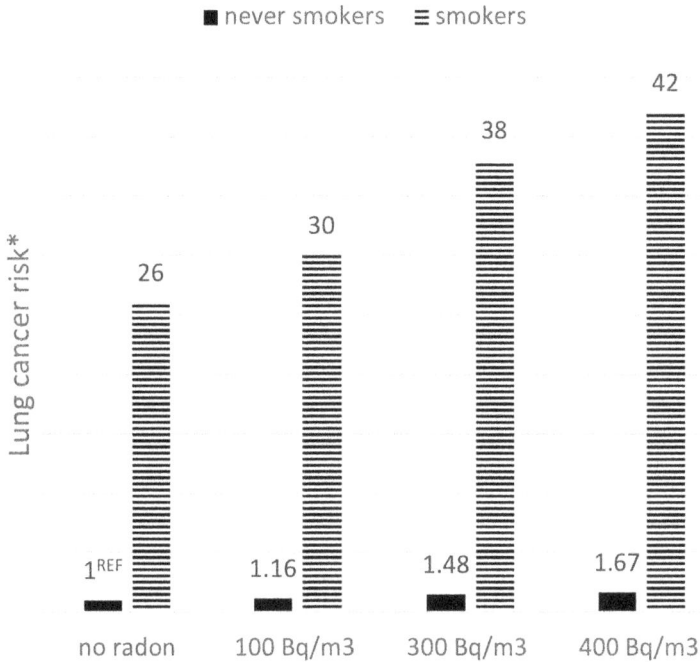

■ never smokers   ☰ smokers

Figure 2.1 Lung cancer risk increases at different radon exposures for smokers and never
  smokers. *Lung cancer risk relative to the reference group; these are never
  smokers without radon exposure that account for a risk of 1 (labelled as 1[REF]).

Source: Figure is based on results from the European collaborative case-control study (Darby
et al., 2006)

on the interaction between radon exposure and secondhand smoke and lung cancer risk.

Radon-attributable mortality

Lung cancer is the cancer with the highest mortality, causing almost one in every
five cancer deaths. Lung cancer has a high incidence due to smoking, exposure
to radon and other modifiable risks factors. Lung cancer survival is very limited;
thus, efforts towards reducing its modifiable risk factors, namely smoking and
radon, are a priority. Noteworthy, radon is considered the first cause of lung cancer among never smokers (World Health Organization, 2009).

The European pooling study estimated that 9% of lung cancer cases in Europe
can be attributed to radon exposure (Darby et al., 2005). Furthermore, numerous

Table 2.2 Studies estimating lung cancer mortality attributable to radon in different countries.

Country, study	Country radon level (Geometric Mean, Bq/m³)	Lung cancer mortality attrib. to radon
Canada (Al-Arydah, 2017)	42	16%
China (Wang et al., 2011)	–	0.2%
France (Ajrouche et al., 2018)	53	10%
Germany (Menzler et al., 2008)	37	5%
Italy (Bochicchio et al., 2013)	52	10%
Mexico (Ángeles & Espinosa, 2015)	83	26%
Portugal (North) (Veloso et al., 2012)	67	25%
Romania (Transilvania) (Truta-Popa et al., 2010)	–	9%
South Korea (Kim et al., 2018)	68	12–25%
Spain (Ruano-Ravina et al., 2021)	43	4%
Switzerland (Menzler et al., 2008)	51	8%
UK (Gray et al., 2009)	14	3%
US (Cao et al., 2017)	25	11%

studies in different countries provided estimations of lung cancer attributable to radon (Table 2.2), with results ranging from 3% (Gray et al., 2009) to 25% (Kim et al., 2018). Results for radon-attributable mortality depend on the radon levels of the country, the smoking prevalence of the population, the definition of radon exposure used and the risk model used for the estimation (Martin-Gisbert et al., 2022).

Notably, most radon-attributable mortality is caused by the joint effect of smoking and low levels of radon exposure. This is due to the overwhelming effect of smoking on lung cancer risk and the fact that most of the population is exposed to low levels of radon (World Health Organization, 2009). For these reasons, smoking cessation would immensely reduce the radon health burden. On the other hand, radon mitigation could prevent one in every four radon-induced deaths (Al-Arydah, 2017; Chen et al., 2012; Kim et al., 2018; Peterson et al., 2013).

Radon exposure in never smokers

Never smokers represent only 15 to 25% of the lung cancer cases (Couraud et al., 2012). Nevertheless, the incidence of lung cancer is so high that even a relatively small fraction of it is still very relevant. In fact, it is so relevant that if lung cancer in never smokers (LCINS) is considered a separate cancer disease, it would be the seventh most common cause of cancer death in the world (Sun et al., 2007).

Evidence of risk factors for LCINS is scarce due to the leading importance of tobacco addiction in producing the lung cancer pandemic. For these reasons, separate studies in never smokers, where tobacco cannot eclipse the effects of other carcinogens, are necessary.

Radon is considered the main risk factor for LCINS (World Health Organization, 2009). A recent meta-analysis comprising 2,341 cases and 8,967 controls observed that lung cancer risk increased 15% for each 100 Bq/m3 increase in radon exposure in never smokers (Cheng et al., 2021). Furthermore, a multicentre case-control study carried out in Spain observed radon exposure in never smokers seemed to be associated with lung cancer cases at an earlier age (Torres-Durán et al., 2015). Finally, a case-control study performed exclusively in never smokers, with more than 500 never smoking lung cancer cases included, found a statistically significant association between radon exposure and LCINS. It was also observed that the risk increase appeared departing from 200 Bq/m3, suggesting that a higher radon concentration is needed for lung cancer to be induced following radon exposure compared to never smokers (Lorenzo-González et al., 2019)

Radon exposure in never smokers

Experimental studies in animals published during the 70s reported lung cancer-induced cases. These experiments were carried out on rats, hamsters and dogs, and they were evaluated and taken into consideration at the IARC Monograph on radon (IARC, 1988). The experimental studies consisted in comparing health outcomes from radon-exposed animals to non-radon-exposed ones or comparing different categories of exposure and different combinations of exposure (including tobacco carcinogens). Radon-exposed animals inhaled extremely high levels of radon (millions of Bq/m3) for a specific time, from weeks to months. The exposed animals showed a significant increase in lung tumours and a dose-response relationship was observed. Furthermore, the combination with other lung carcinogens, such as tobacco smoke, further increased the lung tumour incidence among radon-exposed animals observed (IARC, 1988).

Biokinetic models from ICRP confirm that inhalation of Radon-222 gas is not a relevant source of ionizing radiation itself if compared to its decay products. This is because much of the inhaled Rn^{222} is exhaled back. Coherently, all ^{222}Rn that comes in and out and cannot emit alpha particles inside the body, nor the pulmonary tract, as it did not decay into ^{218}Po during the time it was inside the body. Radon-222 has a half-life of 3.8 days, much longer than that of its alpha-emitting decay products, ^{218}Po (3 minutes) and ^{214}Po (0.000002 seconds), thus would need longer permanence inside the body to produce harmful radiation. Furthermore, Radon-222 is an inert gas; thus, it easily leaves the pulmonary tract through exhalation. On the contrary, ^{218}Po and ^{214}Po are solids that can adhere to the respiratory tract, especially the bronchial epithelium, mainly through diffusion (ICRP, 2018).

Once in the respiratory tract, [218]Po and [214]Po rapidly decay, emitting alpha particles. When an alpha particle impacts a cell nucleus, it causes large DNA damage due to its high levels of energy (Maier et al., 2023).

The genuine molecular pathway in which radon induces carcinogenesis remains unknown. However, different tentative pathways have been studied, including genetic and cytogenetic effects of radon exposure. As such, inconclusive evidence is available on different gene mutations; for instance, mutations in tumour suppressor gene TP53 (Ruano-Ravina et al., 2009), housekeeping gene HPRT (Robertson et al., 2013) or proto-oncogene MET (Gamerith et al., 2022). Chromosome aberrations or reactive oxygen species (ROS) generation have also been studied as tentative radon carcinogenic pathways (Robertson et al., 2013). To date, there is no evidence of a hotspot biomarker for radon-induced tumours.

A further research line is analyzing how radon exposure might have a higher effect on individuals with a higher susceptibility to lung cancer risk. This susceptibility is mostly related to how tobacco is metabolized because some smokers can be affected more than others due to interindividual genetic variability. Some studies have shown that individuals with GSTM1 or GSTT1 genes deleted may have a higher lung cancer risk associated with radon exposure (Bonner et al., 2006; Ruano-Ravina, Pereyra, et al., 2014). Other mechanisms may involve DNA repair genes (Enjo-Barreiro et al., 2022) and a further study has linked radon exposure with genes involved in lung cancer treatment and its response (Ruano-Ravina et al., 2016).

Finally, there is only one study showing an association between lung cancer survival and indoor radon exposure, though this research was performed on never smokers (Casal-Mouriño et al., 2020). More research is clearly needed on this topic.

Radon and diseases other than lung cancer

According to the available evidence to date, radon exposure is a risk factor for lung cancer only. Nevertheless, there is an existing body of evidence, though inconclusive, about the possible causal association between radon and diseases other than lung cancer.

Ecological studies have observed a positive association between radon exposure and stomach cancer (Barbosa-Lorenzo et al., 2017), neurodegenerative disease (Gómez-Anca & Barros-Dios, 2020), brain cancer mortality (Ruano-Ravina, Aragonés, et al., 2017) and esophageal cancer (in men) (Ruano-Ravina, Aragonés, et al., 2014). However, these positive associations have not been further evaluated through adequate epidemiological studies.

According to a recent meta-analysis of available case-control studies, radon is probably not associated with kidney cancer (Chen et al., 2018) nor with cerebrovascular disease (Lu et al., 2022). On the other hand, the evidence available regarding radon exposure and childhood leukemia is inconclusive.

A number of meta-analyses published between 2012 and 2022 informed of a weak yet statistically significant leukemia risk increase among radon-exposed children (Y. Lu et al., 2020; Moon & Yoo, 2021; Ngoc et al., 2022; Tong et al., 2012). However, these meta-analyses have many limitations, including ecological studies (study design that does not allow inference of causality), incoherence in the radon exposure categories and uncertainties in the radon exposure assessments. Remarkably, some of the studies included in the meta-analysis lack radon measurements and use radon estimations instead (Del Risco Kollerud et al., 2014; Demoury et al., 2017; Hauri et al., 2013; Nikkilä et al., 2020; Raaschou-Nielsen et al., 2008). In case-control studies that did include radon measurements, no association between radon exposure and childhood leukemia was found (Cartwright et al., 2002; Kaletsch et al., 1999; Lubin et al., 1998; Steinbuch et al., 1999; Yoshinaga et al., 2005), with the exception of one case-control study in Egypt that reported a positive association (Maged et al., 2000).

Future steps

There is an exceptionally large amount of conclusive evidence of the effects of indoor radon exposure on lung cancer risk, both at high and low radon concentrations. Nevertheless, the undelaying molecular mechanisms involved remain unknown. Radon is the second main cause of lung cancer, yet there are no specific biomarkers to trace whether a lung tumour originates from radon exposure. Unrevealing the molecular pathways for radon carcinogenesis could enable targeted treatments and even targeted early diagnosis.

For other non-pulmonary cancers, there is no conclusive evidence about the potential causative role of the ionizing radiation arising from radon exposure (Barbosa-Lorenzo et al., 2016; Mozzoni et al., 2021). In the case of childhood leukemia, a recent meta-analysis with relevant limitations observed a weak and statistically significant association between radon exposure and leukemia (Moon & Yoo, 2021).

There are other plausible radon effects on health other than cancer that require further study. For instance, ecological studies have observed a potential association between neurodegenerative diseases and radon exposure; however, there are no studies with adequate design evaluating this hypothesis (Gómez-Anca & Barros-Dios, 2020). There is also a potential association between Chronic Obstructive Pulmonary Diseases (COPD) and radon exposure; this association is currently under evaluation through a multicentre case-control study in Spain (Ruano-Ravina et al., 2021).

Reducing radon exposure in both the general population, workers and students, poses a very complex public health problem. It is complex because mitigating its effects clearly requires a multidisciplinary approach involving health professionals, union workers, architects, nuclear safety authorities, departments of work, the general population, councils, regions and many others. The main advantage is the clear evidence available on the health effects arising from radon exposure. Mitigation solutions have also been clearly outlined by different national and international organizations. Radon exposure is perhaps the carcinogenic agent present more frequently in homes and workplaces; thus, addressing this risk is both relevant and challenging. Main priorities include radon prevention in building codes: tackling homes and workplaces (including educational facilities) above reference level and reducing or deleting smoking prevalence.

It is clear that radon causes lung cancer. Unfortunately, lung cancer is a low-survival disease with a high incidence. This health effect is itself so relevant that, from a public health perspective, its mitigation is deemed necessary. Finding out whether radon exposure is, or is not, linked to other diseases would bring further knowledge but likely would lead to a similar solution. To this end, it is time to reduce the exposure to this carcinogen both in the general population and in workers.

Note

1 This work is part of Lucía Martín de Bernardo Gisbert's Doctoral thesis.

References

Ajrouche, R., Roudier, C., Cléro, E., Ielsch, G., Gay, D., Guillevic, J., Micallef, C. M., Vacquier, B., Tertre, A. Le, & Laurier, D. (2018). Quantitative health impact of indoor radon in France. *Radiation and Environmental Biophysics*, *57*(3), 205–214. https://doi.org/10.1007/S00411-018-0741-X

Al-Arydah, M. (2017). Population attributable risk associated with lung cancer induced by residential radon in Canada: Sensitivity to relative risk model and radon probability density function choices: In memory of Professor Jan M. Zielinski. *The Science of the Total Environment*, *596–597*, 331–341. https://doi.org/10.1016/J.SCITOTENV.2017.04.067

Ángeles, A., & Espinosa, G. (2015). Lung cancer mortality from exposure to indoor Radon (222Rn) in Mexico. *Advances in Research*, *5*(3), 1–9. https://doi.org/10.9734/AIR/2015/17736

Auvinen, A., Mäkeläinen, I., Hakama, M., Castrén, O., Pukkala, E., Reisbacka, H., & Rytömaa, T. (1996). Indoor radon exposure and risk of lung cancer: A nested case-control study in Finland. *Journal of the National Cancer Institute*, *88*(14), 966–972. https://doi.org/10.1093/JNCI/88.14.966

Barbosa-Lorenzo, R., Barros-Dios, J. M., Raíces Aldrey, M., Cerdeira Caramés, S., & Ruano-Ravina, A. (2016). Residential radon and cancers other than lung cancer: A cohort study in Galicia, a Spanish radon-prone area. *European Journal of Epidemiology*, *31*(4), 437–441. https://doi.org/10.1007/S10654-016-0134-X

Barbosa-Lorenzo, R., Barros-Dios, J. M., & Ruano-Ravina, A. (2017). Radon and stomach cancer. *International Journal of Epidemiology*, *46*(2), 767–768. https://doi.org/10.1093/IJE/DYX011

Barros-Dios, J. M., Barreiro, M. A., Ruano-Ravina, A., & Figueiras, A. (2002). Exposure to residential radon and lung cancer in Spain: A population-based case-control study. *American Journal of Epidemiology*, *156*(6), 548–555. https://doi.org/10.1093/AJE/KWF070

Barros-Dios, J. M., Ruano-Ravina, A., Pérez-Ríos, M., Castro-Bernárdez, M., Abal-Arca, J., & Tojo-Castro, M. (2012). Residential radon exposure, histologic types, and lung cancer risk. A case—control study in Galicia, Spain. *Cancer Epidemiology Biomarkers and Prevention, 21*(6), 951–958. https://doi.org/10.1158/1055-9965.EPI-12-0146-T/66728/ AM/RESIDENTIAL-RADON-EXPOSURE-HISTOLOGICAL-TYPES-AND

Bochicchio, F., Antignani, S., Venoso, G., & Forastiere, F. (2013). Quantitative evaluation of the lung cancer deaths attributable to residential radon: A simple method and results for all the 21 Italian Regions. *Radiation Measurements, 50*, 121–126. https://doi.org/10.1016/J.RADMEAS.2012.09.011

Bonner, M. R., Bennett, W. P., Xiong, W., Lan, Q., Brownson, R. C., Harris, C. C., Field, R. W., Lubin, J. H., & Alavanja, M. C. R. (2006). Radon, secondhand smoke, glutathione-S-transferase M1 and lung cancer among women. *International Journal of Cancer, 119*(6), 1462–1467. https://doi.org/10.1002/IJC.22002

Cao, X., MacNaughton, P., Laurent, J. C., & Allen, J. G. (2017). Radon-induced lung cancer deaths may be overestimated due to failure to account for confounding by exposure to diesel engine exhaust in BEIR VI miner studies. *PLoS One, 12*(9), e0184298. https://doi.org/10.1371/JOURNAL.PONE.0184298

Cartwright, R. A., Law, G., Roman, E., Gilman, E., Eden, O. B., Mott, M., Muir, K., Goodhead, D., Kendall, G., Doll, R., Day, N., Craft, A., Birch, J. M., McKinney, P. A., Peto, J., Beral, V., Alexander, F. E., Chilvers, C. E. D., Taylor, G. M. . . . Simpson, J. (2002). The United Kingdom childhood cancer study of exposure to domestic sources of ionising radiation: I: Radon gas. *British Journal of Cancer, 86*(11), 1721–1726. https://doi.org/10.1038/SJ.BJC.6600276

Casal-Mouriño, A., Ruano-Ravina, A., Torres-Durán, M., Parente-Lamelas, I., Provencio-Pulla, M., Castro-Añón, O., Vidal-García, I., Pena-Álvarez, C., Abal-Arca, J., Piñeiro-Lamas, M., Fuente-Merino, I., Fernández-Villar, A., Abdulkader, I., Valdés-Cuadrado, L., Barros-Dios, J. M., & Pérez-Ríos, M. (2020). Lung cancer survival in never-smokers and exposure to residential radon: Results of the LCRINS study. *Cancer Letters, 487*, 21–26. https://doi.org/10.1016/J.CANLET.2020.05.022

Chen, B., Yuan, T. W., Wang, A. Q., Zhang, H., Fang, L. J., Wu, Q. Q., Zhang, H. B., Tao, S. S., & Tian, H. L. (2018). Exposure to radon and kidney cancer: A systematic review and meta-analysis of observational epidemiological studies. *Biomedical and Environmental Sciences: BES, 31*(11), 805–815. https://doi.org/10.3967/BES2018.108

Chen, J., Moir, D., & Whyte, J. (2012). Canadian population risk of radon induced lung cancer: A re-assessment based on the recent cross-Canada radon survey. *Radiation Protection Dosimetry, 152*(1–3), 9–13. https://doi.org/10.1093/RPD/NCS147

Cheng, E. S., Egger, S., Hughes, S., Weber, M., Steinberg, J., Rahman, B., Worth, H., Ruano-Ravina, A., Rawstorne, P., & Yu, X. Q. (2021). Systematic review and meta-analysis of residential radon and lung cancer in never-smokers. *European Respiratory Review: An Official Journal of the European Respiratory Society, 30*(159), 1–14. https://doi.org/10.1183/16000617.0230-2020

Council Directive 2013/59/Euratom of 5 December 2013 Laying Down Basic Safety Standards for Protection against the Dangers Arising from Exposure to Ionising Radiation. (2013). https://eur-lex.europa.eu/eli/dir/2013/59/oj

Couraud, S., Zalcman, G., Milleron, B., Morin, F., & Souquet, P. J. (2012). Lung cancer in never smokers—A review. *European Journal of Cancer, 48*(9), 1299–1311. https://doi.org/10.1016/J.EJCA.2012.03.007

Darby, S., Hill, D., Auvinen, A., Barros-Dios, J. M., Baysson, H., Bochicchio, F., Deo, H., Falk, R., Forastiere, F., Hakama, M., Heid, I., Kreienbrock, L., Kreuzer, M., Lagarde, F., Mäkeläinen, I., Muirhead, C., Oberaigner, W., Pershagen, G., Ruano-Ravina, A. . . . Doll, R. (2005). Radon in homes and risk of lung cancer: Collaborative analysis of individual data from 13 European case-control studies. *British Medical Journal, 330*(7485), 223–226. https://doi.org/10.1136/BMJ.38308.477650.63

Darby, S., Hill, D., Deo, H., Auvinen, A., Barros-Dios, J. M., Baysson, H., Bochicchio, F., Falk, R., Farchi, S., Figueiras, A., Hakama, M., Heid, I., Hunter, N., Kreienbrock, L., Kreuzer, M., Lagarde, F., Mäkeläinen, I., Muirhead, C., Oberaigner, W. . . . Doll, R. (2006). Residential radon and lung cancer--detailed results of a collaborative analysis of individual data on 7148 persons with lung cancer and 14,208 persons without lung cancer from 13 epidemiologic studies in Europe. *Scandinavian Journal of Work, Environment & Health, 32 Suppl 1*(1), 1–84. https://pubmed.ncbi.nlm.nih.gov/16538937/

Darby, S., Hill, D., & Doll, R. (2001). Radon: A likely carcinogen at all exposures. *Annals of Oncology, 12*(10), 1341–1351. https://doi.org/10.1023/A:1012518223463

Del Risco Kollerud, R., Blaasaas, K. G., & Claussen, B. (2014). Risk of leukaemia or cancer in the central nervous system among children living in an area with high indoor radon concentrations: Results from a cohort study in Norway. *British Journal of Cancer, 111*(7), 1413–1420. https://doi.org/10.1038/BJC.2014.400

Demoury, C., Marquant, F., Ielsch, G., Goujon, S., Debayle, C., Faure, L., Coste, A., Laurent, O., Guillevic, J., Laurier, D., Hémon, D., & Clavel, J. (2017). Residential exposure to natural background radiation and risk of childhood acute leukemia in France, 1990–2009. *Environmental Health Perspectives, 125*(4), 714–720. https://doi.org/10.1289/EHP296

Enjo-Barreiro, J. R., Ruano-Ravina, A., Pérez-Ríos, M., Kelsey, K., Varela-Lema, L., Torres-Durán, M., Parente-Lamelas, I., Provencio-Pulla, M., Vidal-García, I., Piñeiro-Lamas, M., Fernández-Villar, J. A., & Barros-Dios, J. M. (2022). Radon, tobacco exposure and non-small cell lung cancer risk related to BER and NER genetic polymorphisms. *Archivos de Bronconeumología, 58*(4), 311–322. https://doi.org/10.1016/J.ARBRES.2021.07.006

Environmental Protection Agency. (1987). Radon reference manual. 1987. *Report No.: EPA 520/1–87–20.* Environmental Protection Agency.

Field, R. W., Steck, D. J., Smith, B. J., Brus, C. P., Fisher, E. L., Neuberger, J. S., Platz, C. E., Robinson, R. A., Woolson, R. F., & Lynch, C. F. (2000). Residential radon gas exposure and lung cancer: The Iowa Radon Lung Cancer Study. *American Journal of Epidemiology, 151*(11), 1091–1102. https://doi.org/10.1093/OXFORDJOURNALS.AJE.A010153

Gamerith, G., Kloppenburg, M., Mildner, F., Amann, A., Merkelbach-Bruse, S., Heydt, C., Siemanowski, J., Buettner, R., Fiegl, M., Manzl, C., & Pall, G. (2022). Molecular characteristics of radon associated lung cancer highlights MET alterations. *Cancers, 14*(20). https://doi.org/10.3390/CANCERS14205113

Giraldo-Osorio, A., Ruano-Ravina, A., Varela-Lema, L., Barros-Dios, J. M., & Pérez-Ríos, M. (2020). Residential radon in central and South America: A systematic review. *International Journal of Environmental Research and Public Health, 17*(12), 1–11. https://doi.org/10.3390/IJERPH17124550

Gómez-Anca, S., & Barros-Dios, J. M. (2020). Radon exposure and neurodegenerative disease. *International Journal of Environmental Research and Public Health, 17*(20), 1–17. https://doi.org/10.3390/IJERPH17207439

Gray, A., Read, S., McGale, P., & Darby, S. (2009). Lung cancer deaths from indoor radon and the cost effectiveness and potential of policies to reduce them. *BMJ (Online), 338*(7688), 215–218. https://doi.org/10.1136/BMJ.A3110

Hansen, M. S., Licaj, I., Braaten, T., Lund, E., & Gram, I. T. (2020). The fraction of lung cancer attributable to smoking in the Norwegian Women and Cancer (NOWAC) Study. *British Journal of Cancer, 124*(3), 658–662. https://doi.org/10.1038/s41416-020-01131-w

Hauri, D., Spycher, B., Huss, A., Zimmermann, F., Grotzer, M., von der Weid, N., Weber, D., Spoerri, A., Kuehni, C. E., & Röösli, M. (2013). Domestic radon exposure and risk of childhood cancer: A prospective census-based cohort study. *Environmental Health Perspectives*, *121*(10), 1239–1244. https://doi.org/10.1289/EHP.1306500

IARC. (1988). *IARC monographs on the evaluation of the carcinogenic risks to humans. Volume 43: Man-made mineral fibres and radon*. World Health Organization, International Agency for Research on Cancer.

ICRP. (2014). ICRP Publication 126 radiological protection against radon exposure. *Annals of the ICRP*, *43*(3). https://doi.org/10.1177/ANIB_43_3

ICRP. (2018). ICRP Publication 137 occupational intakes of radionuclides: Part 3. *Annals of the ICRP*, *46*(3–4). http://ani.sagepub.com/

IARC Working Group on the Evaluation of Carcinogenic Risks to Humans., World Health Organization., & International Agency for Research on Cancer. (2004). *Tobacco smoke and involuntary smoking*. IARC Press.

Kaletsch, U., Kaatsch, P., Meinert, R., Schüz, J., Czarwinski, R., & Michaelis, J. (1999). Childhood cancer and residential radon exposure—results of a population-based case-control study in Lower Saxony (Germany). *Radiation and Environmental Biophysics*, *38*(3), 211–215. https://doi.org/10.1007/S004110050158

Kim, S. H., Koh, S. B., Lee, C. M., Kim, C., & Kang, D. R. (2018). Indoor radon and lung cancer: Estimation of attributable risk, disease burden, and effects of mitigation. *Yonsei Medical Journal*, *59*(9), 1123–1130. https://doi.org/10.3349/YMJ.2018.59.9.1123

Krewski, D., Lubin, J. H., Zielinski, J. M., Alavanja, M., Catalan, V. S., Field, R. W., Klotz, J. B., Létourneau, E. G., Lynch, C. F., Lyon, J. I., Sandler, D. P., Schoenberg, J. B., Steck, D. J., Stolwijk, J. A., Weinberg, C., & Wilcox, H. B. (2005). Residential radon and risk of lung cancer: A combined analysis of 7 North American case-control studies. *Epidemiology*, *16*(2), 137–145. https://doi.org/10.1097/01. EDE.0000152522.80261.E3

Létourneau, E. G., Krewski, D., Choi, N. W., Goddard, M. J., Mcgregor, R. G., Zielinski, J. M., & Du, J. (1994). Case-control study of residential radon and lung cancer in Winnipeg, Manitoba, Canada. *American Journal of Epidemiology*, *140*(4), 310–322. https://doi.org/10.1093/OXFORDJOURNALS.AJE.A117253

Lorenzo-González, M., Ruano-Ravina, A., Torres-Durán, M., Kelsey, K. T., Provencio, M., Parente-Lamelas, I., Leiro-Fernández, V., Vidal-García, I., Castro-Añón, O., Martínez, C., Golpe-Gómez, A., Zapata-Cachafeiro, M., Piñeiro-Lamas, M., Pérez-Ríos, M., Abal-Arca, J., Montero-Martínez, C., Fernández-Villar, A., & Barros-Dios, J. M. (2019). Lung cancer and residential radon in never-smokers: A pooling study in the Northwest of Spain. *Environmental Research*, *172*, 713–718. https://doi.org/10.1016/J. ENVRES.2019.03.011

Lorenzo-Gonzalez, M., Ruano-Ravina, A., Torres-Duran, M., Kelsey, K. T., Provencio, M., Parente-Lamelas, I., Piñeiro-Lamas, M., Varela-Lema, L., Perez-Rios, M., Fernandez-Villar, A., & Barros-Dios, J. M. (2020). Lung cancer risk and residential radon exposure: A pooling of case-control studies in northwestern Spain. *Environmental Research*, *189*. https://doi.org/10.1016/J.ENVRES.2020.109968

Lu, L., Zhang, Y., Chen, C., Field, R. W., & Kahe, K. (2022). Radon exposure and risk of cerebrovascular disease: A systematic review and meta-analysis in occupational and general population studies. *Environmental Science and Pollution Research International*, *29*(30), 45031–45043. https://doi.org/10.1007/S11356-022-20241-X

Lu, Y., Liu, L., Chen, Q., Wei, J., Cao, G., & Zhang, J. (2020). Domestic radon exposure and risk of childhood leukemia: A meta-analysis. *Journal of B.U.ON. : Official Journal of the Balkan Union of Oncology*, *25*(2), 1035–1041. https://pubmed.ncbi.nlm.nih. gov/32521903/

Lubin, J. H., Linet, M. S., Boice Jr., J. D., Buckley, J., Conrath, S. M., Hatch, E. E., Kleinerman, R. A., Tarone, R. E., Wacholder, S., & Robison, L. L. (1998). Case-control study of childhood acute lymphoblastic leukemia and residential radon exposure. *Journal of the National Cancer Institute, 90*(4), 294–300. https://doi.org/10.1093/jnci/90.4.294

Lubin, J. H., Wang, Z. Y., Boice, J. D., Xu, Z. Y., Blot, W. J., De Wang, L., & Kleinerman, R. A. (2004). Risk of lung cancer and residential radon in China: Pooled results of two studies. *International Journal of Cancer, 109*(1), 132–137. https://doi.org/10.1002/IJC.11683

Maged, A. F., Mokhtar, G. M., El-Tobgui, M. M., Gabbr, A. A., Attia, N. I., & Shady, M. M. A. (2000). Domestic radon concentration and childhood cancer study in Cairo, Egypt. *Journal of Environmental Science & Health Part C, 18*(2), 153–170. https://doi.org/10.1080/10590500009373519

Maier, A., Bailey, T., Hinrichs, A., Lerchl, S., Newman, R. T., Fournier, C., & Vandevoorde, C. (2023). Experimental setups for in vitro studies on radon exposure in mammalian cells-A critical overview. *International Journal of Environmental Research and Public Health, 20*(9). https://doi.org/10.3390/IJERPH20095670

Martin-Gisbert, L., Ruano-Ravina, A., Varela-Lema, L., Penabad, M., Giraldo-Osorio, A., Candal-Pedreira, C., Rey-Brandariz, J., Mourino, N., & Pérez-Ríos, M. (2022). Lung cancer mortality attributable to residential radon: A systematic scoping review. *Journal of Exposure Science & Environmental Epidemiology.* https://doi.org/10.1038/S41370-022-00506-W

Menzler, S., Piller, G., Gruson, M., Rosario, A. S., Wichmann, H. E., & Kreienbrock, L. (2008). Population attributable fraction for lung cancer due to residential radon in Switzerland and Germany. *Health Physics, 95*(2), 179–189. https://doi.org/10.1097/01.HP.0000309769.55126.03

Moon, J., & Yoo, H. K. (2021). Residential radon exposure and leukemia: A meta-analysis and dose-response meta-analyses for ecological, case-control, and cohort studies. *Environmental Research, 202.* https://doi.org/10.1016/J.ENVRES.2021.111714

Mozzoni, P., Pinelli, S., Corradi, M., Ranzieri, S., Cavallo, D., & Poli, D. (2021). Environmental/occupational exposure to radon and non-pulmonary neoplasm risk: A review of epidemiologic evidence. *International Journal of Environmental Research and Public Health, 18*(19). https://doi.org/10.3390/IJERPH181910466

National Research Council. (1999). *Health effects of exposure to radon: BEIR VI.* National Academies Press. https://doi.org/10.17226/5499

Ngoc, L. T. N., Park, D., & Lee, Y. C. (2022). Human health impacts of residential radon exposure: Updated systematic review and meta-analysis of case-control studies. *International Journal of Environmental Research and Public Health, 20*(1). https://doi.org/10.3390/IJERPH20010097

Nikkilä, A., Arvela, H., Mehtonen, J., Raitanen, J., Heinäniemi, M., Lohi, O., & Auvinen, A. (2020). Predicting residential radon concentrations in Finland: Model development, validation, and application to childhood leukemia. *Scandinavian Journal of Work, Environment & Health, 46*(3), 278–292. https://doi.org/10.5271/sjweh.3867

Parkin, D. M. (2011). Tobacco-attributable cancer burden in the UK in 2010. *British Journal of Cancer, 105,* S6–S13. https://doi.org/10.1038/BJC.2011.475

Peterson, E., Aker, A., Kim, J., Li, Y., Brand, K., & Copes, R. (2013). Lung cancer risk from radon in Ontario, Canada: How many lung cancers can we prevent? *Cancer Causes and Control, 24*(11), 2013–2020. https://doi.org/10.1007/S10552-013-0278-X

Raaschou-Nielsen, O., Andersen, C. E., Andersen, H. P., Gravesen, P., Lind, M., Schüz, J., & Ulbak, K. (2008). Domestic radon and childhood cancer in Denmark. *Epidemiology, 19*(4), 536–543. https://doi.org/10.1097/01.ede.0000288431.93533.7f

Radon Causes Most of the Radiation Received by Finns | Säteilyturvakeskus (STUK). (n.d.). Retrieved October 10, 2023, from www.sttinfo.fi/tiedote/69879452/radon-causes-most-of-the-radiation-received-by-finns?publisherId=64456131

Radon map | Environmental Protection Agency. (n.d.). Retrieved October 13, 2023, from www.epa.ie/environment-and-you/radon/radon-map/

Robertson, A., Allen, J., Laney, R., & Curnow, A. (2013). The cellular and molecular carcinogenic effects of radon exposure: A review. *International Journal of Molecular Sciences*, *14*(7), 14024–14063. https://doi.org/10.3390/IJMS140714024

Ruano-Ravina, A., Aragonés, N., Kelsey, K. T., Pérez-Ríos, M., Piñeiro-Lamas, M., López-Abente, G., & Barros-Dios, J. M. (2017). Residential radon exposure and brain cancer: An ecological study in a radon prone area (Galicia, Spain). *Scientific Reports*, *7*(1). https://doi.org/10.1038/S41598-017-03938-9

Ruano-Ravina, A., Aragonés, N., Pérez-Ríos, M., López-Abente, G., & Barros-Dios, J. M. (2014). Residential radon exposure and esophageal cancer. An ecological study from an area with high indoor radon concentration (Galicia, Spain). *International Journal of Radiation Biology*, *90*(4), 299–305. https://doi.org/10.3109/09553002.20 14.886792

Ruano-Ravina, A., Cameselle-Lago, C., Torres-Durán, M., Pando-Sandoval, A., Dacal-Quintas, R., Valdés-Cuadrado, L., Hernández-Hernández, J., Consuegra-Vanegas, A., Tenes-Mayén, J. A., Varela-Lema, L., Fernández-Villar, A., Barros-Dios, J. M., & Pérez-Ríos, M. (2021). Indoor radon exposure and COPD, synergic association? A multicentric, hospital-based case-control study in a radon-prone area. *Archivos de Bronconeumologia*, *57*(10), 630–636. https://doi.org/10.1016/J.ARBR.2020.11.020

Ruano-Ravina, A., Faraldo-Vallés, M. J., & Barros-Dios, J. M. (2009). Is there a specific mutation of p53 gene due to radon exposure? A systematic review. *International Journal of Radiation Biology*, *85*(7), 614–621. https://doi.org/10.1080/09553000902954504

Ruano-Ravina, A., Kelsey, K. T., Fernández-Villar, A., & Barros-Dios, J. M. (2017). Action levels for indoor radon: Different risks for the same lung carcinogen? *European Respiratory Journal*, *50*(5). https://doi.org/10.1183/13993003.01609-2017

Ruano-Ravina, A., Pereyra, M. F., Castro, M. T., Pérez-Ríos, M., Abal-Arca, J., & Barros-Dios, J. M. (2014). Genetic susceptibility, residential radon, and lung cancer in a radon prone area. *Journal of Thoracic Oncology: Official Publication of the International Association for the Study of Lung Cancer*, *9*(8), 1073–1080. https://doi. org/10.1097/JTO.0000000000000205

Ruano-Ravina, A., Torres-Durán, M., Kelsey, K. T., Parente-Lamelas, I., Leiro-Fernández, V., Abdulkader, I., Abal-Arca, J., Montero-Martínez, C., Vidal-García, I., Amenedo, M., Castro-Añón, O., Golpe-Gómez, A., González-Barcala, J., Martínez, C., Guzmán-Taveras, R., Provencio, M., Mejuto-Martí, M. J., Fernández-Villar, A., & Barros-Dios, J. M. (2016). Residential radon, EGFR mutations and ALK alterations in never-smoking lung cancer cases. *The European Respiratory Journal*, *48*(5), 1462–1470. https://doi.org/10.1183/13993003.00407-2016

Ruano-Ravina, A., & Wakeford, R. (2020). The increasing exposure of the global population to ionizing radiation. *Epidemiology (Cambridge, Mass.)*, *31*(2), 155–159. https:// doi.org/10.1097/EDE.0000000000001148

Steinbuch, M., Weinberg, C. R., Buckley, J. D., Robison, L. L., & Sandler, D. P. (1999). Indoor residential radon exposure and risk of childhood acute myeloid leukaemia. *British Journal of Cancer*, *81*(5), 900–906. https://doi.org/10.1038/sj.bjc.6690784

Sun, S., Schiller, J. H., & Gazdar, A. F. (2007). Lung cancer in never smokers—a different disease. *Nature Reviews Cancer*, *7*(10), 778–790. https://doi.org/10.1038/nrc2190

Tong, J., Qin, L., Cao, Y., Li, J., Zhang, J., Nie, J., & An, Y. (2012). Environmental radon exposure and childhood leukemia. *Journal of Toxicology and Environmental Health. Part B, Critical Reviews*, *15*(5), 332–347. https://doi.org/10.1080/10937404 .2012.689555

Torres-Durán, M., Ruano-Ravina, A., Parente-Lamelas, I., Leiro-Fernández, V., Abal-Arca, J., Montero-Martínez, C., Pena-Álvarez, C., Castro-Añón, O., Golpe-Gómez, A., Martínez, C., Guzmán-Taveras, R., Mejuto-Martí, M. J., Provencio, M., Fernández-Villar,

A., & Barros-Dios, J. M. (2015). Residential radon and lung cancer characteristics in never smokers. *International Journal of Radiation Biology, 91*(8), 605–610. https://doi.org/10.3109/09553002.2015.1047985

Truta-Popa, L. A., Dinu, A., Dicu, T., Szacsvai, K., Cosma, C., & Hofmann, W. (2010). Preliminary lung cancer risk assessment of exposure to radon progeny for transylvania, Romania. *Health Physics, 99*(3), 301–307. https://doi.org/10.1097/HP.0B013E3181C03CDE

UK radon. (n.d.). *UKradon—What is radon?* October 6, 2023. www.ukradon.org/information/whatisradon#:~:text=The radioactive elements formed by, can lead to lung cancer.

Veloso, B., Nogueira, J. R., & Cardoso, M. F. (2012). Lung cancer and indoor radon exposure in the north of Portugal—An ecological study. *Cancer Epidemiology, 36*(1), e26–e32. https://doi.org/10.1016/J.CANEP.2011.10.005

Wang, J. B., Fan, Y. G., Jiang, Y., Li, P., Xiao, H. J., Chen, W. Q., Wei, W. Q., Zhou, Q. H., Qiao, Y. L., & Boffetta, P. (2011). Attributable causes of lung cancer incidence and mortality in China. *Thoracic Cancer, 2*(4), 156–163. https://doi.org/10.1111/J.1759-7714.2011.00067.X

World Health Organization. (2009). *WHO handbook on indoor radon: A public health perspective*. World Health Organization.

Yoshinaga, S., Tokonami, S., Akiba, S., Nitta, H., & Kabuto, M. (2005). Case-control study of residential radon and childhood leukemia in Japan: Results from preliminary analyses. *International Congress Series, 1276*, 233–235. https://doi.org/10.1016/J.ICS.2004.09.050

3 Radon and citizen science

*Meritxell Martell, Tanja Perko, Sylvain Andresz,
Caroline Schieber, Yevgeniya Tomkiv,
Alison Dowdall, Leo McKittrick and
Veronika Oláhné Groma*

Introduction

Radon awareness-raising campaigns have been one of the main communication tools used to inform citizens about the risks of this indoor air pollutant (Hevey et al., 2023). The European Basic Safety Standards Directive (BSS) 2013/59/ EURATOM also requires European Member States to develop "strategies for communication to increase public awareness and inform local decision makers, employers and employees of the risk of radon, including in relation to smoking" as part of their National Action Plans to address the long-term risks from radon exposure (Perko et al., 2023). In this regard, most of the European Member States have embarked on national radon awareness campaigns to reduce exposure. However, risk communication on radon has mostly focused on testing rather than on mitigation (Mc Laughlin et al., 2022; Hevey et al., 2023; Apers et al., 2023).

Research indicates that residents in areas with high radon presence who are aware of its harmful effects may not be concerned about living in a house with elevated radon levels, resulting in a lack of radon testing and mitigation efforts (Poortinga et al., 2011; Hevey, 2015; Lofstedt, 2019). Perko and Turcanu (2020) and Apers et al. (2023) note that higher levels of stakeholder engagement in radiological protection and, in particular, in protection from radon, are needed to address the value action gap observed in radon risk mitigation and increase protective behaviour concerning indoor radon. Citizen science might be one of the tools for engaging the public in scientific research whilst generating scientific data, enhancing knowledge and positively changing their radiation protection behaviour (Santori et al., 2021; Chase & Levine, 2018).

A limited number of CS projects exist in the realm of radiation protection (Kenens et al., 2020), and in the context of radon, their presence is even scarcer (Martell et al., 2021). Furthermore, the existing CS projects in the field of radon are not registered in CS databases, which makes it even more difficult to identify past or ongoing CS initiatives in this area. The few CS projects in the field of radon have mostly concentrated on radon testing to the neglect of mitigation (ibid.). Citizen science has proven effective in promoting radon testing,

DOI: 10.4324/9781032618180-4

enhancing understanding of radon and boosting confidence in the ability of citizen scientists to contact a radon mitigation professional. However, it did not necessarily lead to hiring a mitigation professional or a belief that radon mitigation would effectively reduce the threat of radon exposure (Stanifer et al., 2022).

The EURATOM Horizon 2020 project RadoNorm has made an important contribution by supporting CS projects in the field of radon and registering them in several CS databases, in particular EU-Citizen.Science and SciStarter. RadoNorm aims to manage risks from radon and naturally occurring radioactive materials (NORM) exposure situations in order to assure effective radiation protection based on improved scientific evidence and social considerations (Kulka et al., 2022). One of the objectives of the RadoNorm project is to establish a CS incubator for radon priority areas and a network of CS projects to address radon testing and mitigation across Europe. In this regard, RadoNorm partners from France, Hungary, Ireland and Norway developed and tested different models as pilot projects for CS to manage radon risk over a period of six months. This chapter focuses on discussing the potential of CS in radon testing and mitigation, drawing insights from lessons learned in the four RadoNorm CS pilot projects.

Theoretical background

The term CS evolved in the 1990s and became renowned globally in 2012 due to the increased rise in the number of funding schemes, projects and publications (Vohland et al., 2021). Citizen science is a growing phenomenon worldwide and broadly refers to the active participation of citizens in scientific research as researchers, offering non-scientists an opportunity to collaborate with scientists and participate in all stages of the research process (Land-Zandstra et al., 2021), thereby bridging the wide gap between science and society. CS initiatives also have the potential to increase citizens' trust in scientific outcomes.

Even though citizen scientists have predominantly been involved in data collection and analysis, different citizen science typologies have been developed depending on the level of citizen participation in scientific research. In RadoNorm, we adopt the typologies proposed by Haklay's (2013), which goes from level 1 "crowdsourcing" to level 4 "extreme citizen science," as described in Martell et al. (2021). Table 3.1 presents the levels of participation applied in the RadoNorm citizen science pilot projects. All four CS projects involved higher levels of citizen participation, with France, Hungary and Ireland utilising participatory science and Norway utilising the extreme citizen science approach.

Method

The RadoNorm project proposed a model of CS project in the field of radon and a set of indicators to assess the contribution of a project to radon research based on a scoping review of the relevant project in the database, systematic

Table 3.1 Levels of citizen participation applied to the RadoNorm citizen science pilot projects.

Levels according to Haklay (2013)		RadoNorm citizen science pilot projects
Level 3	Participatory science	In **France**, participants contributed to improving an existing online self-diagnostic tool that helps to identify entry and transfer mechanisms in a building and offer mitigation solutions.
		In **Hungary**, high school students in three institutions developed a toolkit consisting of various low-cost measurement sensors to measure air quality, including radon.
		In **Ireland**, citizens contributed to co-create a Do It Yourself (DIY) "toolkit" for mitigation.
Level 4	Extreme citizen science	In **Norway**, citizens were approached to define the aim of the pilot citizen science project themselves, the research question and their level of involvement.

Source: Modified from Martell et al. (2022).

review of scientific publications and consultations and interview with radon experts and practitioners as described in Martell et al. (2021). From these, each RadoNorm partner defined the aims and objectives of their pilot project, the collaborating partners and roles, the recruitment strategies and alignment with ECSA 10 principles, which all proved to be different. The main characteristics of the four RadoNorm citizen science pilot projects are described in what follows.

(1) France (Table 3.2). The focus of the project was the 'radon building diagnosis', which is supposed to take place after (high) radon concentrations are measured and prior to mitigation itself. However, this diagnosis is hardly implemented in France. To remedy this situation, the pilot project recruited citizens already aware of radon from Pays Vesoul Val-de-Saône (East of France) to (a) test an existing online self-evaluation guide for radon diagnosis (the guide was the offspring of a former research project funded by Interreg 2017–2019), (b) report on their operational experience through survey and (c) meet with radon/building experts. Citizens contributed to improvements in both the form and content of the guide and to ensure that it would be better fit for purpose. Comparison of the guide with experts' practices in the field provided additional perspectives on building diagnosis. The conclusions from the project were ascertained during a plenary videoconference, and the effects of the project on both citizens and experts were surveyed formally (Andresz et al., 2023).

Table 3.2 Roles of the partners in the RadoNorm pilot CS project in France.

Partners	Role in the pilot CS project
Nuclear Protection Evaluation Centre (CEPN)	Leader; developed the protocol and associated documents; analysed and compiled the data; disseminated the results.
High School of Engineering and Architecture Freiburg (HEIA, Switzerland)*	Provided expertise on radon/building, participated in the meetings; disseminated the results.
Centre of Studies and Experience in Risk, Environment, Mobility and Urbanism (CEREMA, France)*	Provided expertise on radon/building; participated in the meetings; disseminated the results.
Pays de Vesoul Val de Saône (PVVS)	Assisted with the recruitment of citizens and local organisations; disseminated the results.
Citizens	Tested the guide; reported on their operational experience; contributed to the evolutions.

Source: Authors' own elaboration.

*Initial developers of the guide.

(2) Hungary (Table 3.3). The aim of the RadoNorm pilot CS project in Hungary was to test whether it is feasible to develop an affordable toolkit measuring several air quality and radiation components, including radon, CO_2, particulate matter and CO. This project is driven by the need to address the limited public awareness regarding radon, particularly its correlation with indoor air pollutants and the multiple beneficial effects of mitigation.

Unlike the other three pilot projects, which mostly featured adult citizens, the pilot project in Hungary focused on high school students as citizen scientists. Consequently, workshops were first organised to contact teachers from three high schools in Budapest and Székesfehérvár respectively, and then for secondary school students of 14 to 18 years old, where interested students were invited to become citizen scientists. The 18 natural science students involved developed a toolkit and tested it through a series of short-term individual measurements at different locations at schools and residences. The results were analysed and interpreted in collaboration with teachers and researchers. The pilot CS project included an interdisciplinary component, as students also developed the technical documentation of the toolkit in easy-to-understand English in collaboration with students of design and communication, who accompanied the development and testing process in addition to helping in the dissemination tasks. The project is disseminated on a student-created website.

(3) Ireland (Table 3.4). The CS pilot project in Ireland aimed to establish a Do-it-Yourself (DIY) toolkit to increase the rate of mitigation. The project was an opportunity for citizens and radon experts to develop a DIY

Table 3.3 Roles of the partners in the RadoNorm pilot CS project in Hungary.

Partners	Role in the pilot CS project
Centre for Energy Research (EK)	Leader; conceptualised and led the assembly process of the tool.
Schools—students	Volunteers aged 14 to 18 participated in this project-based extracurricular course. Their areas of interest are natural science and communication. The first group developed, set up and tested the toolkit whilst the second group wrote the English documentation of the toolkit and designed the logo and website for dissemination.
Schools—teachers	Ensured the conditions of the thematic course; took part in the development work; supervised the students in collaboration with researchers.
Radiation and air quality measurement experts and development engineers	Advised on device selection and contributed to the technical design of the toolkit.

Source: Authors' own elaboration.

Table 3.4 Roles of the partners in the RadoNorm pilot CS project in Ireland.

Partners	Role in the pilot CS project
EPA	Leader; developed a toolkit with the input of citizen scientists and a radon mitigation contractor; co-ordinated the project locally between the citizen scientists, libraries and radon contractor; reported on the participant testing of the toolkit and finalised it based on the user experience; provided passive radon detectors for participants to carry out pre and post mitigation measurements.
Residents (owners of dwellings)	Take part in a workshop to co-design the DIY toolkit; test the DIY toolkit in their homes; participate in two meetings and, after testing the toolkit, provide feedback on how it can be improved.
Radon contractor	A radon mitigation contractor registered as such with the EPA. The contractor supported the citizen scientists by providing advice on mitigation; assisted with the mitigation work as required; and worked on the development of the toolkit with EPA and citizen scientists before and after it was used.
Wexford local authorities	Supported engagement on radon with residents in Wexford County.
Wexford local libraries staff	Administered the loan scheme of digital radon monitors; facilitated public meetings for CS recruitment and workshop to co-design toolkit. Provided contact details and completed expression of interest and/or consent forms from residents who had borrowed the digital monitors and observed high readings to EPA.
Healthy Wexford	Established contacts with their existing networks within the community; promoted the CS project through their social media.

Source: Author's own elaboration.

approach for the mitigation of private dwellings and to build on the partner-ship between EPA, the radon contractor, Wexford local authorities, local libraries and Healthy Wexford.

The rationale for this project was the knowledge that despite finding high levels of radon after carrying out a radon test, only one in five householders would carry out mitigation work to reduce these levels (Dowdall et al., 2016). County Wexford in Ireland is a radon priority area where EPA data shows that approximately 500 homes have been tested above 200 Bq/m^3, and 400 of these have not yet been remediated. The main reasons given for not remediating are cost and lack of concern. Householder knowledge of radon mitigation "may not translate into behaviour in the face of psycho-social and financial barriers to action" (EPA, 2019). Using a DIY approach could be a lower cost mitigation for some citizens.

The CS pilot project focused on citizens who had borrowed digital monitors from any of the five Wexford libraries that have loan schemes and had observed high levels of radon. Following a public meeting, New Ross Library (which pro-vided information on the project) attendees were invited to express their interest in taking part in the co-design and testing of the DIY mitigation toolkit and to carry out a three-month radon test. The citizen scientists were recruited once the three-month test confirmed that their levels of radon were above the 200 Bq/m3 reference level. Their role included taking part in a workshop to co-design the DIY toolkit, testing the DIY toolkit in their homes and, after testing the toolkit, providimg feedback on how it can be improved. The input of citizen scientists in this CS project was used to generate new knowledge and understanding for the EPA and mitigation experts on support for householders to help them remediate.

(4) Norway (Table 3.5). The Norwegian pilot CS initiative attempted to come as close as possible to 'extreme citizen science' by letting participants define the project themselves and choose the extent to which they wanted to be involved in the actions identified to improve mitigation. The Norway Pilot CS project took place in Gjøvik municipality, one of the radon priority areas, in collaboration with the inter-municipal public service company Miljørettet helsevern IKS and the City Lab Bylab in Gjøvik. The pilot project began with a public workshop at the local library, where citizens openly discussed barriers to mitigation and what could be done to get more people to miti-gate. Based on the results of the discussion, citizens, together with project partners, decided that the pilot project will focus on improved public infor-mation about radon, radon detection and radon mitigation; better access to radon testing kits; information about professionals on radon mitigation and about costs of mitigation; and information about DIY options for radon mitigation.

Table 3.5 Roles of the partners in the RadoNorm pilot CS project in Norway.

Partners	Role in the pilot CS project
NMBU/CERAD	Lead organiser; contact point for the information of the project; analysed the CS project; disseminated the project through the communication channels of the university (press releases, social media, etc.).
Gjøvik municipality	Assisted with the dissemination of information about the project for the purpose of participant recruitment. Local library hosted the first project workshop.
Miljørettet helsevern IKS intermunicipal public service company	Supported engagement of residents for the RadoNorm citizen science project; assisted with dissemination, recruitment and practical organisation locally.
Residents (individuals)	Defined and shared the CS project and co-created an information package about radon measurement and mitigation for the local population tailored to answer questions and concerns from the residents.

Source: Authors' own elaboration.

Results

Conceptualisation of the CS project

As shown in Table 3.1, the RadoNorm CS pilot projects were designed to fit in levels 3 and 4 of Haklay's (2013) typology. Thus, Andresz et al. (2023) argue that the CS project in France differentiated itself from former projects as citizens were at the core of the project in the different steps: in the data collection (the whole radon diagnosis guide was submitted to their analysis), meetings (the words of the citizens and the experts were equal) and overall organisation (the development of the project and the conclusions were entirely based on the citizens' answers and proposals) blurring the usual delineation between experts and citizen. In Hungary, the CS project students not only technically assembled a measurement device but also designed the logo, created the website and helped develop the technical documentation of the toolkit in easy-to-understand English. The project in Ireland involved citizens in the development of the DIY toolkit by using it in their own homes. Their input generated new knowledge and understanding for the EPA and mitigation experts on what support householders need to help them remediate. Finally, in the "extreme" citizen science model used in Norway, citizens themselves defined the barriers to mitigation the project should focus on, co-created a radon guide, recruited more participants for the measurement campaign and designed a public information meeting.

Recruitment and engagement of citizen scientists

Finding citizens to volunteer to take part in a CS project on radon proved challenging in all four pilot projects. The complexity of the topic and the fact that the start of the projects coincided with the Covid-19 pandemic made recruitment even more challenging.

The CS projects used a range of methods to recruit citizen scientists, from flyers, posters, announcements and articles in local newspapers, radio interviews, specific webpages of the project, press releases, personal letters, social media channels of the partners, emails and phone calls. In some cases, like in Norway, using the established social media channels of the local authorities (i.e. Facebook) had a much bigger impact on recruitment than the physical flyers.

The number of citizen scientists mobilised and engaged in the four RadoNorm pilot projects is shown in Table 3.6. However, it should be borne in mind that the total number of citizens is not necessarily a measure of the success of the CS pilot project.

Andresz et al. (2023) provide possible (combining) explanations for the low number of participants in the case of France: an opportunistic targeted sampling in a small size target, a small team of experts, possible erosion of the interest of the inhabitants about radon, the complexity of the topic and cognitive/emotional response to radon.

In Hungary, the cooperation with schools was undertaken at different intensities, and the RadoNorm partners adapted to the motivation and possibilities of the teachers and students. Nevertheless, the high level of motivation of the participants has resulted in planning further development, and the project is still ongoing in the new educational course 2023–2024 with extensions and improved technical solutions.

In Ireland, whilst 40 people attended the first public meeting, 26 attendees completed a form expressing an interest in taking part in the RadoNorm CS pilot project and agreed to carry out a free three-month radon test in their homes. Nineteen of these completed the test and received a report with their result. Of

Table 3.6 Number of citizen scientists involved in the RadoNorm CS pilot projects (without supporting personnel such as teachers or contractors).

Country	France	Hungary	Ireland	Norway
Total n° of citizen scientists	6*	18	9/17**	100+***

Source: Authors' own elaboration.

*In France, the target set by the project leader was 20 citizen scientists.

**26 potential participants agreed to carry out a three-month-test following a public meeting, but 19 completed the test. Nine participants had radon levels above the 200 Bq/m³ reference level and were, therefore, the target for the CS project.

***Eight participants took part in the first workshop, 97 participated in the measurement campaign and 32 in the final meeting of the CS project.

these, nine had radon levels above the 200 Bq/m^3 reference level, and these were recruited as citizen scientists to co-design the DIY toolkit.

In Norway, the first meeting was attended by five local residents, three representatives from Miljørettet helsevern IKS and three researchers from NMBU, although two additional residents asked to be involved but could not attend that first meeting. In the second online meeting, six residents, three of whom were new, were involved and two from Miljørettet helsevern IKS participated. Free radon dosimeters were offered to 100 households owning and residing in certain target areas of Gjøvik, and 97 households signed up for the campaign. A total of 32 participants took part in the final meeting.

The motivations for participating in the different CS projects were varied. Some citizen scientists had a general interest in the radon problem, some were interested in the technical and professional aspects, and some had questions about radon and radon mitigation. The motivations for teachers in Hungary to get involved in the project were linked to the innovative work with students and the resulting publications which are necessary for their career advancement. In France, the motivation of being involved in a European project was appealing to some participants as well as interest in the topic. In Ireland, free expert advice and a DIY mitigation kit were a good motivation to become involved in the project, as some had been engaged in the topic of radon over many years but had not been motivated to mitigate. In Norway, some citizen scientists were motivated by the little attention paid to radon in the Gjøvik area and the uncertainty regarding radon being a health problem as one of the reasons for being involved in the project.

Outcomes

In all four countries, the citizen science project developed guidelines, supporting materials or toolkits related to increasing information about radon in general or, more specifically, about radon testing and mitigation. These materials and DIY toolkits were co-created by the participants through different workshops and meetings and, in some cases, feedback was provided on how to improve them. In the case of France, where the guide already existed, the modifications proposed by citizens helped to improve it and confirmed its usefulness with regard to fostering action in the radon post-measurement step. The project will help to increase the visibility of this guide.

In the case of Ireland, an increase in the rate of mitigation was observed in the CS pilot project. At 66%, this is significantly higher than the rate of mitigation of 22% reported by Dowdall et al. (2016). However, it should be noted that this rate of mitigation is based on private householders who paid for both testing and mitigation in their homes, whereas for the participants in the citizen science pilot project, there were no costs incurred and they received tailored support from the radon contractor.

Discussion and conclusion

The RadoNorm CS pilot projects yielded valuable insights into critical considerations when establishing CS projects in radon protection. It became evident that a well-defined project concept, including clear objectives and the potential for active citizen scientist engagement, should be established from the project's inception. Setting achievable goals that contribute to radon measurement and mitigation is paramount. Another crucial lesson learned pertains to the openness of CS projects, adherence to open science principles and the careful handling of data protection and personal/sensitive data and ethical principles. The project implementation plan must ensure that planned activities are attainable for citizen scientists while also emphasising an effective strategy to engage citizen scientists. Furthermore, assessing the potential impact of the CS project, including its behavioural and socio-cultural effects on participating citizen scientists and their increased knowledge of radon, is essential. Lastly, assembling a knowledgeable and skilled team with experience in radon-related research activities and managing citizen science projects is vital for CS project success.

Role of authorities and researchers in promoting citizen science initiatives on radon

Martell et al. (2021) assessed that the CS initiatives on radon were promoted top-down by researchers and public authorities, unlike CS initiatives in other fields, such as those in the environmental field. Similarly, the RadoNorm CS pilot projects were launched and led either by national authorities (Ireland), research institutes (Hungary) or universities (France and Norway). However, all these organisations collaborated closely with several local or national partners, which was crucial in recruiting participants, disseminating the projects through different channels or providing technical expertise. The financial support to promote these initiatives has come from the European project RadoNorm. Drawing from the valuable lessons learned from the pilot projects, RadoNorm opened a call for proposals in November 2022 to fund citizen science projects in Europe for a six-month period. Out of 19 proposals, six projects are funded in Italy, Poland, Portugal, Slovakia, Slovenia and Spain. The promoters of these CS projects include NGOs, universities, national and local authorities, private persons, research institutes and, in most cases, consortiums that promote and run these initiatives in their regions. These projects make use of different citizen science approaches and focus on indoor radon measurements and mitigation (Martell & Perko, 2023).

Radon awareness and awareness of citizen science

Despite the short time frame of the CS pilot projects—the duration of the projects was a maximum of six months, although in some cases, the time frame was shorter (less than five months in France)—it was found that the level of radon awareness and awareness of citizen science increased. The four RadoNorm CS pilot projects raised citizen awareness about radon risks and mitigation actions to reduce exposure. They also created spaces to talk with experts on the topic of radon. At the same time, experts became aware of the opportunities that citizen science may offer in the area of radon management. Experts increased their awareness of the existence of the knowledge and perception gaps between them and citizens and realised the need to facilitate mutual learning for all parties involved. Nevertheless, there is a need for science-based surveys designed to collect data and empirical evidence to find out why people test (or not) and mitigate (or not) in different social environments and cultures, as well as the relationship between radon awareness and protection behaviour of residents under radon risk. RadoNorm makes a contribution in this regard by implementing a European RadoNorm behavioural atlas (Perko et al., 2021) in collaboration with radon authorities of 16 European Member States. This survey also explores the motivation behind individuals' interest in participating in potential citizen science projects related to radon.

Challenges of radon CS projects

One of the challenges in the RadoNorm CS projects is the fact that radon management is a complex topic, which makes it difficult for citizens to be motivated to engage. In addition, radon is an indoor pollutant and, therefore, a household problem which might not have the same potential for community engagement among stakeholders compared to other environmental issues. In Ireland, by prioritising citizens who had previously borrowed digital radon monitors from libraries, the organisers acknowledged two potential limitations: first, they primarily invited library members as participants, and second, they excluded individuals with no prior awareness of radon.

Similar to other CS projects (National Academies of Sciences, Engineering and Medicine, 2018), the majority of participants in Norway were retired males who typically have more free time compared to other target participants with more diverse ethnic, age or cultural backgrounds, such as young homeowners or families with young children.

Implementing effective CS projects proved challenging in all cases due to the significant time required. For instance, one of the planned activities in Norway was to develop videos on DIY mitigation solutions, which did not happen due to time constraints. In Ireland, a limitation arose from the time commitment required of citizen scientists to use a DIY approach to mitigation and carry out one-on-one interactions with the radon contractor. The skill set and the tools necessary for mitigation work were also identified as limiting factors.

CS for radon mitigation

The RadoNorm CS pilot projects highlighted the need to have access to and consultations with mitigation or building professionals before making any decisions. In addition, the need to offer simple mitigation solutions was also pointed out. After all, the low number of professionals interested in undertaking mitigation work, the uncertainty regarding the effectiveness and the real costs are recognised as important barriers to domestic radon mitigation (Hevey et al., 2023).

In line with the findings of Stanifer et al. (2022), our study also underscores the necessity for additional investigations to comprehensively explore the role of citizen science in addressing home radon mitigation.

Acknowledgements

We would like to sincerely thank all citizen scientists who invested time in the four RadoNorm CS projects; this study would not have been possible without their support.

Funding information

The work described in this chapter was conducted within the RadoNorm project. The RadoNorm project has received funding from the Euratom research and training programme 2019–2020 under grant agreement No. 900009. In the case of Norway, the study received funding from the Research Council of Norway, grant No. 313070 and 223268. This publication reflects only the authors' views. Responsibility for the information and views expressed therein lies entirely with the authors. The European Commission is not responsible for any use that may be made of the information it contains.

References

Andresz, S., Marchand-Moury, A., Goyette-Pernot, J., Rivière, A., & Schieber, C. (2023). When citizen science meets radon building diagnosis: Synthesis of a French pilot project developed in the framework of the European RadoNorm research project [version 2; peer review: 1 approved, 1 approved with reservations]. *Open Research Europe*, *3*(106), 1–24. https://doi.org/10.12688/openreseurope.15968.2

Apers, S., Vandebosch, H., & Perko, T. (2023). Clearing the air: A systematic review of mass media campaigns to increase indoor radon testing and mitigation. *Communications*. https://doi.org/10.1515/commun-2021-0141

Chase, S. K., & Levine, A. (2018). Citizen science: Exploring the potential of natural resource monitoring programs to influence environmental attitudes and behaviors. *Conservation Letters*, *11*(2), e12382.

Council Directive 2013/59/EURATOM. (2014). *Basic safety standards for protection against the dangers arising from exposure to ionising radiation, and repealing Directives 89/618/Euratom, 90/641/Euratom, 96/29/Euratom, 97/43/Euratom and 2003/122/Euratom, C. D. 2013/59/EURATOM, 2013.* http://eur-lex.europa.eu/legal-content/EN/TXT/PDF/?uri=CELEX:32013L0059&from=en

Dowdall, A., Fenton, D., & Rafferty, B. (2016). The rate of radon remediation in Ireland 2011–2015: Establishing a base line rate for Ireland's National Radon Control Strategy. *Journal of Environmental Radioactivity*, *162*, 107–112. https://doi.org/10.1016/j.jenvrad.2016.05.001

EPA. (2019) National radon control strategy. Phase 1 (2014–2018). *Final report*. www.epa. ie/publications/monitoring--assessment/radon/national-radon-control-strategy-year-4-report-to-government.php

Haklay, M. (2013). Citizen science and volunteered geographic information: Overview and typology of participation. In D. Sui, S. Elwood, & M. Goodchild (Eds.), *Crowdsourcing geographic knowledge* (pp. 105–122). Springer. https://doi. org/10.1007/978-94-007-4587-2_7

Hevey, D. (2015) Review of public information programmes to enhance home radon screening uptake and home mitigation. In *EPA Report No 170*. Environmental Protection Agency.

Hevey, D., Perko, T., Martell, M., Bradley, G., Apers, S., & Rovenská, K. N. (2023). A psycho-social-environmental lens on radon air pollutant: Authorities', mitigation contractors', and residents' perceptions of barriers and facilitators to domestic radon mitigation. *Frontiers in Public Health*, *11*, 1–13. https://doi.org/10.3389/ fpubh.2023.1252804

Kenens, J., Van Oudheusden, M., Yoshizawa, G., & Van Hoyweghen, I. (2020). Science by, with and for citizens: Rethinking 'citizen science' after the 2011 Fukushima disaster. *Palgrave Communications*, *6*, Art. No. 58, 1–8. https://doi.org/10.1057/ s41599-020-0434-3

Kulka, U., Birschwilks, M., Fevrier, L., Madas, B., Salomaa, S., Froňka, A., Perko, T., Wojcik, A., & Železnik, N. (2022). RadoNorm—towards effective radiation protection based on improved scientific evidence and social considerations—focus on RADON and NORM, *EPJ N: Nuclear Science Technologies*, *8*(38), 1–13. https://doi. org/10.1051/epjn/2022031

Land-Zandstra, A., Agnello, G., & Gültekin, Y. S. (2021). Participants in citizen science. In K. Vohland et al. (Eds.), *The science of citizen science* (pp. 243–259). Springer. https://doi.org/10.1007/978-3-030-58278-4_13

Lofstedt, R. (2019). The communication of radon risk in Sweden: Where are we and where are we going?. *Journal of Risk Research*, *22*(6), 773–781.

Martell, M., Andresz, S., Dowall, A., Fenton, D., Olahne Groma, V., Kenens, J., McKittrick, L., Perko, T., Tomkiv, Y., & Schieber, C. (2022). *Citizen science model for radon prone areas (citizen science incubator), D6.9*. RadoNorm.

Martell, M., & Perko, T. (2023). Selection of citizen science projects through open call. *Milestone 85 RadoNorm Project*.

Martell, M., Perko, T., Tomkiv, Y., Long, S., Dowdall, A., & Kenens, J. (2021). Evaluation of citizen science contributions to radon research. *Journal of Environmental Radioactivity*, *237*(106685), 1–10. https://doi.org/10.1016/j.jenvrad.2021.106685

Mc Laughlin, J. P., Gutierrez-Villanueva, J., & Perko, T. (2022). Suggestions for improvements in national radon control strategies of member states which were developed as a requirement of EU directive 2013/59 EURATOM. *International Journal of Environmental Research and Public Health*, *19*(7), 1–8. https://doi.org/10.3390/ ijerph19073805

National Academies of Sciences, Engineering, and Medicine. (2018). *Learning through citizen science: Enhancing opportunities by design*. The National Academies Press. https://doi.org/10.17226/25183

Perko, T., Martell, M., Rovenská, K. N., Fojtiková, I., Paridaens, J., & Geysmans, R. (2023). *National radon action plans: Results of the EU-RAP study's review and evaluation, Radiation protection EC, Luxembourg*. European Comission.

Perko, T., & Turcanu, C. (2020). Is internet a missed opportunity? Evaluating radon websites from a stakeholder engagement perspective. *Journal of Environmental Radioactivity*, *212*, 106123, https://doi.org/10.1016/j.jenvrad.2019.106123.

Perko, T., Turcanu, C., Hoti, F., Thijssen, P., & Muric, M. (2021). *RadoNorm pilot study report from public opinion survey, Belgium 2020–2021 Development of a modular questionnaire for investigating societal aspects of radon and NORM.* https://doi.org/10.20348/STOREDB/1174/1251.

Poortinga, W., Bronstering, K., & Lannon, S. (2011). Awareness and perceptions of the risks of exposure to indoor radon: A population-based approach to evaluate a radon awareness and testing campaign in England and Wales. *Risk Analysis, 31*(11), 1800–1812. www.ncbi.nlm.nih.gov/pubmed/21477087

Santori, C., Keith, R. J., Whittington, C. M., Thompson, M. B., Van Dyke, J. U., & Spencer, R. -J. (2021). Changes in participant behaviour and attitudes are associated with knowledge and skills gained by using a turtle conservation citizen science app. *People and Nature, 2021*(3), 66–76. https://doi.org/10.1002/pan3.10184

Stanifer, S., Hoover, A. G., Rademacher, K., Rayens, M. K., Haneberg, W., & Hahn, E. J. (2022). Citizen science approach to home radon testing, environmental health literacy and efficacy. *Citizen Science: Theory and Practice, 7*(1), 26, pp. 1–13. https://doi.org/10.5334/cstp.472

Vohland, K., Göbel, C., Balázs, B., Butkevičienė, E., Daskolia, M., Duží, B., Hecker, S., Manzoni, M., & Schade, S. (2021). Citizen science in Europe. In K. Vohland et al. (Ed.), *The science of citizen science* (pp. 45–53). Springer. https://doi.org/10.1007/978-3-030-58278-4_3

Part II

Communicating public health

4 Public health communication from digital native media

Carmen Costa-Sánchez, Sofia Gomes and Xosé López-García

Introduction

The transformation of the digital ecosystem in the third millennium has not only led to convergence processes in the communication industry but has also enabled the introduction of many new communication channels, which have multiplied the offer and fed the superabundant communication flows.

The continuous interplay of old and new media in shaping the network society soon established a renewed hybrid model. This has been a recurring theme throughout media history, but now, in the context of the current century, it represents a novel aspect of the digital landscape (Chadwick, 2013). This new digital scenario involves a coexistence and competition between legacy media and new digital native media, and both are confronted with the disruptive force of well-established social media, leading to significant challenges to traditional media models (Campos-Freire, 2008).

One of the consequences, after intense years of digital disruption and successive changes in the communication landscape and the strategy of different players, has been the media industry's growing concern about the communication impact of "tech giants" steadily climbing positions in the communication ecosystem, and the consequences of the disruption of media business models (Newman, 2019).

Technological disruptions and increased competition in the digital media landscape have transformed the market conditions for news media, raising numerous concerns about the future of journalism (Westlund et al., 2021), an ongoing issue to this day.

Changes in the communication ecosystem have enhanced processes, motivated numerous debates and multiplied their complexity. While it is true that the digitisation of the media has enabled changes in the production and consumption of news (Kramp & Loosen, 2018), the rise of the social web and the consolidation of social media have expanded and multiplied the possibilities for active audiences to participate in journalism. These processes have led to the emergence of terms such as participatory journalism (Singer et al., 2011) or

DOI: 10.4324/9781032618180-6

network journalism (Heinrich, 2011), among many others specific to the new scenario. Meanwhile, new models of public interest journalism have developed in recent years, with renewed funding models (Carvajal et al., 2012) and a wide variety of initiatives, both in form and the entities and organisations promoting them. This context, typical of the network society, has fuelled a set of factors that have altered the communication dynamics between journalists and audiences in the digital environment. Today, audiences play a prominent role in all communication strategies emphasising greater co-creation and participation. Their voices and contributions are now accorded more significance compared to the past.

The new challenges require journalism to find novel approaches to balancing participation and journalistic values (Hujanen, 2016). This entails effectively combining the emotional charge of information impact of information to engage in the new media landscape (Beckett & Deuze, 2016) while fostering empathy, a crucial quality that modern digital journalists must establish with their users (Gluck, 2016). These dimensions, among others, are evolving and seem to be consolidating. All available data point to the fact that the future of journalism is online (Van der Haak et al., 2012). However, this certainty alone is not enough; societal transformations and the media ecosystem call for re-evaluating digital journalism. It must now face new challenges in a scenario shaped by artificial intelligence and next-generation technologies in order to fulfil its roles in the new social, political and economic context (Zelizer, 2017) and with the conviction shown by the main communication stakeholders that digital journalism is more than digital technology because it is journalism of today that gives technology purpose, form, perspective and meaning, and not the other way around (Zelizer, 2019).

Journalism is in society. It belongs to society and must be embedded in a specific social, political and economic context, in which it must face the challenges that secure its future.

Native media

The new challenges of journalism in the third decade of the third millennium are a test for all stakeholders in the field of communication, but especially for professionals and legacy media, which have made the digital transition, and digital native media, born and designed for the networked communication scene.

Digital native media, as channels with structures and techniques naturally adapted to the digital environment, are key players in the digital communication ecosystem. Many of the initiatives that have emerged in recent years in the field of public health communication are in line with this typology of digital native media, which has become an emerging phenomenon expanding worldwide and is currently a central axis for crisis communication and the implementation of risk communication strategies. In the third decade of the current millennium, there is a rich and diverse ecosystem of new native media, which will foreseeably add many more initiatives in the coming years, a significant number of which will strive to innovate within the field of digital communication.

Digital journalism has been around for almost three decades now. The data suggest that after more than 26 years of research in digital journalism, the analysis of the progress made confirms that digital journalism is a well-established and developing discipline (Salaverría, 2019), with important challenges ahead of it in this decade. Digital journalism studies, driven by social science perspectives, need, among other things, to reinforce the connections between empirical research and the many conceptual discussions that dominate the journalism field (Steensen et al., 2019). A way to address the knowledge of important transformations since the late 1990s, when online newspapers had a technical infrastructure, incipient organisational and communication patterns and a set of products very different from those of printed newspapers (Boczkowski, 2005). At this stage of transition between the existing and the emerging, several works analysed the processes of change and transformation of the old media—press, radio and television—through metamorphosis (Fidler, 1997) and the features of the new media—ubiquity, hypertextuality, multimedia, interactivity and instantaneity— and the implications for journalists and audiences (Pavlik, 2001). From the onset, different types of online media and the first effects of the World Wide Web on the journalistic profession and culture were noticed, hand in hand with multimedia journalism (Deuze, 2004), which introduced important novelties in the formal construction of news pieces.

Since the first half of the millennium's first decade, changes have taken place rapidly, and multimedia journalism has evolved alongside the emergence of new media. According to the perception of professionals and experts in the field of communication, it is safe to say that when we talk about new media, we are referring to an alternative media ecosystem to the traditional one, which innovates, uses new narrative formats and has a new relationship with the audience (Cabrera Méndez et al., 2019). These new media, apart from imitating more or less the legacy media, evolved and shaped renewed structures that tried to anticipate the future of networked communication with the elements that defined the communication of new players (blogs, social media, etc.). Interactivity, multimedia and hypertextuality were the foundations of the new model. Still, in a short time, the social web and the new formats fed a great diversity of expressive modalities, formats and strategies. In a scenario of mutual influences, the different players explored new territories, searched for sustainability models and fostered specialisation. The changes were dizzying, and innovation was imposed in the hope of finding solutions to the current challenges of journalism (García-Avilés, 2021) and evolving in the construction of other possible journalisms.

Digital native media are here today because they have emerged and consolidated their positions in recent years. They will be here in the future because the future is digital and requires new media designed for this changing and constantly evolving landscape. The rise of digital native media in recent years confirms this forecast, but it does not mean that all those emerging will be consolidated. Many have already fallen along the way in recent years of the 21st century due to their structural issues and the limitations they presented (Salaverría, 2020), but despite the challenges in finding sustainable models, the expansion of digital native media is unquestionable (Salaverría, 2021). Securing a successful future for these digital native media, regardless of their content or specialised themes, entails facing the challenge of diversifying their funding sources and establishing sustainable business models (Tejedor et al., 2020). There is little doubt among researchers that these media, born within the digital ecosystem, have great potential to improve journalism (Buschow, 2020). Despite facing various limitations, they actively strive to do so within a market dominated by large platforms (Rietveld & Schilling, 2021) and a society increasingly reliant on such platforms (van Dick et al., 2018). This leads to important changes and conflicts.

Media specialised in public health

The concept of "health" is complex, difficult to define and has undergone a significant evolution since its inception, as what we consider health problems today are not the same as those identified by our ancestors, and situations that years ago were considered illnesses are not so today. The most classic definition, prevailing until the end of the 19th century, defined health as the absence of disease, thus phrasing the concept negatively. In the mid-20th century, the World Health Organisation (WHO) proposed a positive definition that involved various dimensions of the person and not only the biological dimension.

The concept of health as the absence of disease is the central idea of what is known as the "biomedical model", which postulates as basic ideas that all health problems have a biological origin and that body and mind function independently. Biological reductionism and mind-body dualism are the starting premises. Hence, diagnosis and treatment focused on the person's physical aspects are the only possible actions within the health system. On the other hand, there is the "biopsychosocial model" (proposed by Engel, 1977), in which health is conceived as a whole in which biological, psychological and social aspects converge. Scott Ratzan, a researcher acknowledged for his contribution to the field of Health Communication, also argues that health is an essential component of civil society, so its social context must be considered. An individual's health encompasses several meanings, namely technical, moral or philosophical (Gomes, 2020; Ratzan, 2002; Naidoo & Wills, 1998).

Although most financial resources in today's health systems are allocated to health care, they are not the most important determinant of health. According to the Lalonde system, lifestyle factors are the most important determinants of individual and collective health status in developed countries (Colomer & Álvarez-Dardet, 2006).

Therefore, health communication activities, as defined by different authors (Alcalay, 1999; Ishikawa et al., 2010; Schiavo, 2007), are very important for changing habits and behaviours that are detrimental to our health. So, who are the agents in charge of Health Communication? In essence, it is possible to identify the following types of actors involved:

1. Health professionals: The professionals who take care of our health are the most direct reference when it comes to giving advice, guidelines and instructions not only when we are unwell but also on a preventive basis. For example, in a community health crisis, public institutions play a very important role in preventing and mitigating disease through messages encouraging people to adopt safe behaviours (Vos & Buckner, 2016).
2. Internet and social media as sources of information: Users have become accustomed to using the Internet to access health information (Observatorio Nacional de las Telecomunicaciones y de la Sociedad de la Información, 2020), which is why the use of the Internet as a source of health information is growing every day (Rodríguez-González, 2021). The consolidation of the use of digital platforms has created a hybrid media system in which new and old media coexist (Casero-Ripollés, 2020).
3. From eating habits to specific symptoms, different sources on the Internet disseminate health-related content. The controversy revolves around the credibility of these sources, which is why different accreditation systems are being developed for websites and apps so that citizens can trust this information (Fernández Silano, 2014; Sánchez-Bocanegra & Sánchez-Laguna, 2012). In social media, the chaos is even greater, and the emergence of leaders in the 2.0 environment who talk about health without training highlights a complex problem that only seems to have a solution through citizen literacy.
4. Educational organisations: Schools, institutes, universities, research centres . . . Health Communication is part of health education (Jover Ibarra, 2006). In this sense, the institutions responsible for educating citizens from an early age are also committed to promoting healthy habits.
5. Healthcare organisations: Hospitals, primary care centres, health councils, the Ministry of Health, the World Health Organisation, pharmaceutical companies, research institutions, patient or family associations, foundations related to this field, etc. These entities, at a local (micro) or global (macro) level, can convey messages to society, encouraging lifestyles that are more beneficial to our health.
6. Media: With the information they publish, the media can help to disseminate and promote healthy lifestyles, they can publicise diseases and diseases that are socially invisible, etc. In particular, regular programmes contribute to disseminating health-related content that may interest the population. In the press, some

Spanish and Portuguese newspapers have created a dedicated space for health news. In contrast, others still include this information in the Society area, in a mishmash of topics that undermine its weight and importance in the news at large.

The native media represent a new initiative for disseminating specialised information, which has proved to be a thematic area of special interest to citizens (Catálan Matamoros, 2019).

The focus on the quality of the health information that reaches the citizens has been a constant in the literature (Revuelta-De-la-Poza, 2019).

The attention and time devoted to health information coverage has grown exponentially. The activities of the press offices of public health institutions and others to prepare and send information to the media to reach the population have also grown. The relevance and immediacy of this information have prompted the joint work of these two public players in a global humanitarian effort (Ruão et al., 2020).

However, the pandemic has also shown that, due to the high levels of misinformation circulating, the limited scientific culture of society and the degree of health information specialisation, journalism must dedicate human and material resources to the coverage of information in this thematic area. The specialised digital native media are also a channel for disseminating specialised information specific to the network. However, in the Sánchez-García and Amoedo-Casais (2021) picture, Health, Religion and Gastronomy account for only 1.7% of the overall digital media.

Some selected case studies

An intentional case selection of digital native media has been made, from which the authors offer a brief overview in the form of a concise analysis sheet representing noteworthy initiatives within the Latin American sphere.

Salud con Lupa

Salud con Lupa is a digital platform dedicated to public health in Peru and Latin America that belongs to the Asociación de Periodismo con Lupa. The purpose of the Latin American project, focusing on the prevailing misinformation and the relationship between health and other domains and spheres of economic and political power, is self-defined as follows: "we strive to ensure our news helps improve the quality of public debate and the authorities' decision-making process. Above all, we aspire to make them part of the citizens' everyday conversations". Furthermore, "we want people to talk more about health and realise that it is an issue that involves many interests and is highly related to economics and politics at local and global levels".

This digital native media has a main menu with ten sub-sections:

- News. From a global perspective, the News section covers information on epidemics, health policy and generally urgent or crisis issues in the health field.
- Check it out. This is the section dedicated to combating misinformation. With questions related to hoaxes (e.g. 'Does intermittent fasting help you lose weight?') or with specific topics (e.g. 'Three ways to measure how fit you are, without focusing on your weight'), it publishes checks on false news circulating on the web and helps disseminate healthy lifestyle habits with understandable and interesting information.
- Mental health. This is a specialised area aimed at disseminating content of interest on mental health. It is a clear commitment within the native medium to highlight the importance of mental health in the health of citizens.
- Climatopedia. The climate becomes the protagonist of this section of the content's menu. Under a one health approach, the native medium thus circumscribes the importance of climatic phenomena for health.
- Environment. Under this premise of interconnection between human health and that of the ecosystems, the contents concerning the environment, the energy issue and environmental protection are presented as a section of their own.
- Gender. Another novel section in a specialised health media. Gender equality is treated as the main content, as well as male violence or sexual and reproductive health, promoting a broader culture of the concept of health and specifically focusing on women.
- Tabletop. Specialising in nutrition, yet another area of interest in health information, this section focuses on healthy eating habits and features analytical articles related to food.
- Opinion, Interviews and Specials. With different types of texts according to the journalistic genre (opinion articles, interviews and special pieces).

It allows a subscription to a newsletter to keep up to date with the latest news. It has an extensive social media outreach strategy, adapting to the language and format of each social media. It is part of *The Global Investigative Journalism Network*, founded in 2003, and at its core, GIJN comprises an international association of non-profit journalism organisations.

ConSalud.es

This online media is specialised in health and is published by the Spanish publishing group *Mediforum*, which also edits other publications. It has been running for ten years (2013 to 2023). It focuses primarily on health management and the stakeholders in this field, featuring a broad range of content grouped under the following sections:

- Policy. The Health Policy section includes coverage of the Ministerio de Sanidad (Ministry of Health), Consejerías (Regional Councils), Administration (Administration), Parlamentos (Parliaments) and La trastienda del Ministerio (The backroom of the Ministry). This section covers mainly institutional sources with competencies in Spain in the field of health. It includes statements of institutional officials, legislative initiatives, public budgets, etc.
- Industry. This section includes other private stakeholders impacting the health sector in Spain, namely pharmaceuticals, insurance companies, private healthcare and innovation-related issues.
- Technology. This subsection is specifically dedicated to technological developments such as artificial intelligence or news from the industry. It includes the journal *SaluDigital* (Tu Revista de e-health; Your e-health journal), a specialised online journal published by the same group.
- Professionals. In this section, the goal is to meet the needs of the professional sector through a wide range of topics, including all types of health professionals (physicians, nurses, psychologists, pharmacists, dentists, etc.). Professionals are also given a voice through the sub-section Conversations with C, a platform for video interviews with healthcare professionals.
- Autonomous Regions. Dividing the territorial organisation of Spain into autonomous regions, this section covers all of them to provide information on the latest developments in the medical sector and relevant health policy.
- Patients. Addresses the cure of diseases and advances in the treatments of different diseases, including thematic channels on specific topics: Chronocity Channel (for chronic diseases), Coronavirus Special, HIV Channel and Health Education Channel.
- Opinion. Opinion articles on current health issues.
- ConsaludTV. From a small set, it features monographs with videos on different topics and the participation of healthcare professionals.
- ConSaludPodcast. In a podcast format, it offers three series of fully edited podcasts available from the YouTube channel.

It is worth noting that many of the contents are recycled across the sections, for although they are divided into different themes, the boundaries between them are flexible, allowing many pieces to fit into different sections.

On the other hand, ConSalud.es also has a newsletter to which readers can subscribe to receive the latest news. They are present on all social media (X, Facebook, Instagram, LinkedIn and YouTube, and they announce that soon users can also join their WhatsApp groups).

Saúde Online

Saúde Online is an informative media in health, mainly addressed to health professionals and society at large. Its contents are produced by accredited journalists who abide by the rights and duties provided for in the Constitution of the Republic, as well as in the Portuguese Press Law and the Portuguese Journalists' Statute.

Following the same structure as other information media, Saúde Online is subdivided into the following sections:

* Specials. An area dedicated to treating informative content generated from topics such as awards, congresses and commemorative events in the field of health.
* National. Section with news, interviews and opinion articles about the Portuguese scene.
* World. Similar to the previous section, it contains news, interviews and opinion articles covering the international scene (outside Portugal).
* Research. This section provides news on new scientific advances, particularly articles about studies conducted in Portugal and abroad.
* Symposiums. A section mainly devoted to testimonies from expert sources. Many of these experts' contributions are recorded in conferences, symposia or seminars.
* Specialities. This website section is subdivided into different medical specialities. The name of each category suggests a speciality: allergy, cardio, derma, pain, endo, gastro, gynaecology, haematology, hypertension and CVR, infectious, internal medicine, FGM, neuro, onco, ophthalmology, paediatrics, psychiatry, pneumology, rheuma and uro.

Each of these subcategories includes articles in their respective speciality. The main focus is on information about scientific research in the field. For each of the specialities, there is an agenda with upcoming events, a section for interviews with specialists in the area and an area for opinion articles. It is worth noting that some content is cross-referenced between the different sections. For example, the same interview can be available in the section "National" and the section "Symposiums".

Saúde Online has a link to *Saúde OnlineTV*, which is not highlighted on the website's homepage. The media outlet's website also dedicates a section to the *Saúde Notícias* journal, where all the publication's editions are available in open access. Because its target audience is health professionals, *Saúde Online* is linked to microsites with restricted access through a login. Moreover, this media outlet is on social media (Facebook, X, Instagram, LinkedIn and YouTube) and has an app.

Minha Vida

This Brazilian media outlet was launched in 2006. Since then, it has shared content on health, beauty, nutrition and quality of life. This content is produced by more than a dozen specialised journalists and more than a thousand contributors in the medical field. The website is divided into the following categories:

- Health. This section is subdivided into general health, examinations, contraception, weight loss, immunity, medicinal plants, sleep and vaccinations. Each subcategory provides topical articles on the themes.
- Food. This area is subdivided into general food, drinks, "nutrients", "diet plan" and "supplements". Each subcategory contains news or articles with suggestions for recipes and diet plans.
- Beauty. This category focuses on beauty-related topics: hair, body, make-up, face and nails. It also provides articles on aesthetic procedures.
- Fitness. This section is divided into general fitness, sports and exercises. It is an area dedicated to articles on sports and exercises.
- Family. This category is dedicated to the family and topics related to pregnancy and childhood, such as breastfeeding, children's nutrition, maternity and fertility.
- Well-being. An area covering general well-being, behaviour, home care, relationships, sex and addictions.
- Special channels. This section has more specific and temporary contents: exercise guide, health in colour, my life as a mother, appointments and special blood donation.
- Specialists. An area to introduce the professionals who collaborate with *Minha Vida* and its contents.

Minha Vida is present on social media (Facebook, X, YouTube, Pinterest and Instagram) and provides a newsletter on request. It also has a section for consultation, which users can access and where registered doctors can answer (identifying themselves) any queries they may have.

Conclusions

Although it is not a major specialisation, digital native media have taken off their activity on the web in the field of health in the Latin American context. Thus, these initiatives focus on citizens or health professionals, and that determines their content strategies and approaches to the topics covered.

Within the media studied, two are mainly technical and specialised (*Consalud.es* and *Saudeonline*), and the other two are mostly informative and aimed at society at large (*Saludconlupa.com* and *Minhavida*). There are important differences between the last two, as the former maintains a critical-analytical focus, while the latter is linked to health and lifestyle, beauty and well-being.

Table 4.1 Proposed classification of digital native media specialising in health.

Scope	Typology	Cases
Digital native media and health	Specialised: Targeted primarily at health professionals as the preferred audience Informative: aimed at society as the preferred audience. • *Informative-Analytical* • *Informative-Lifestyle*	• Consalud • Saudeonline • Saludconlupa • Minhavida

Source: Authors' own elaboration.

Hence, diversity in digital native media specialising in health is an intrinsic characteristic of the sector that has not yet been explored in previous relevant research. Nevertheless, Table 4.1 suggests the classification of digital native health media in the Latin American sphere.

Furthermore, all the analysed media have made a considerable effort to establish a presence across various digital platforms, including social media. They also produce newsletters to build reader loyalty and leverage audiovisual platforms (such as television channels) to enhance the appeal of their content. Digital native media specialising in health demonstrate commitment to adapting to the participatory environment with push and pull initiatives (Lopezosa et al., 2020), ensuring their audiences appreciate their content.

Given the growing misinformation (López-García et al., 2021), which poses serious health risks, the private and public sectors should be committed to reliable initiatives that provide citizens with verified and relevant online health information. Considering the crisis we are still facing today and possible new public health crises (Costa-Sánchez & López-García, 2020), citizens have the right to be informed with quality and rigour and on the platforms through which they access information regularly (Costa-Sánchez & Míguez-González, 2018). Because only by having health information (health literacy) can preventive and responsible behaviours be adopted, thus promoting the health of citizens (Gomes, 2019).

References

Alcalay, R. (1999). La comunicación para la salud como disciplina en las universidades estadounidenses. *Public Health*, *5*(3), 192–195. https://scielosp.org/pdf/rpsp/v5n3/top192.pdf

Beckett, C., & Deuze, M. (2016). On the role of emotion in the future of journalism. *Social Media+Society*, *2*(3), 1–6. http://journals.sagepub.com/doi/pdf/10.1177/2056305116662395

Boczkowski, P. J. (2005). *Digitizing the news: Innovation in online newspapers*. The MIT Press.

Buschow, C. (2020). Why do digital native news media fail? An investigation of failure in the early start-up phase. *Media and Communication*, *8*(2). https://doi.org/10.17645/mac.v8i2.2677

Cabrera Méndez, M., Codina, L., & Salaverría-Aliaga, R. (2019). Qué son y qué no son los nuevos medios. 70 visiones de expertos hispanos. *Revista Latina de Comunicación Social, 74*, 1506–1520. https://doi.org/10.4185/RLCS-2019-1396

Campos-Freire, F. (2008). Las redes sociales trastocan los modelos de los medios de comunicación tradicionales. *Revista Latina de Comunicación Social, 63*, 287–293. https://doi.org/10.4185/RLCS-63-2008-767-287-293

Carvajal, M., García-Avilés, J. A., & González, J. L. (2012). Crowdfunding and non-profit Media. *Journalism Practice, 6*, 638–647. https://doi.org/10.1080/1751278 6.2012.667267

Casero-Ripollés, A. (2020). Impact of COVID-19 on the media system. Communicative and democratic consequences of news consumption during the outbreak. *El Profesional de la Información, 29*(2), e290223.

Catálan Matamoros, D. (2019). *Communication and public health challenges in Europe. Specialised journalism, sources and media coverage in times of anti-vaccine lobby* [Ph.D. Thesis, University of the Basque Country].

Chadwick, A. (2013). *The hybrid media system: Politics and power.* Oxford University Press.

Colomer, C., & Álvarez-Dardet, C. (2006). *Promoción de la salud y cambio social.* Editorial Masson.

Costa-Sánchez, C., & López-García, X. (2020). Comunicación y crisis del coronavirus en España: Primeras lecciones. *Profesional de la Información, 29*(3), e290304. https://doi.org/10.3145/epi.2020.may.04

Costa-Sánchez, C., & Míguez-González, M. I. (2018). Use of social media for health education and corporate communication of hospitals. *Profesional de la información, 27*(5), 1145–1154. https://doi.org/10.3145/epi.2018.sep.18

Deuze, M. (2004). What is multimedia journalism? *Journalism Studies, 5*(2), 139–152. https://doi.org/10.1080/1461670042000211131

Engel, G. (1977). The need for a new medical model: A challenge for biomedicine. *Science, 196*, 129–136. https://doi.org/10.1126/science.847460

Fernández Silano, M. (2014). La Salud 2.0 y la atención de la salud en la era digital. *Revista médica de Risaralda, 20*(1), 41–46.

Fidler, R. (1997). *Mediamorphosis: Understanding new media.* Pine Forge Press.

García-Avilés, J. A. (2021). Review article: Journalism innovation research, a diverse and flourishing field (2000–2020). *Profesional de la información, 30*(1), e300110.

Gluck, A. (2016). What makes a good journalist? Empathy as a central resource in journalistic work practice. *Journalism Studies, 17*(7), 893–903. www.tandfonline.com/doi/full/10.1080/1461670X.2016.1175315

Gomes, E. S. (2019). *Jornalismo e prevenção em saúde: Retratos da imprensa portuguesa entre 2012 e 2014* [Tese de Doutoramento, Universidade do Minho].

Gomes, E. S. (2020). O Jornalismo em saúde e as fontes de informação: o caso da COVID-19 em Portugal. *Revista De La Asociación Española De Investigación De La Comunicación, 7*(14), 127–149. https://doi.org/10.24137/raeic.7.14.6

Heinrich, A. (2011). *Journalistic practice in interactive spheres.* Routletge.

Hujanen, J. (2016). Participation and the blurring values of journalism. *Journalism Studies, 17*(7), 871–880. www.tandfonline.com/doi/full/10.1080/1461670X.2016.1171164

Ishikawa, H., & Kiuchi, T. (2010). Health literacy and health communication. *BioPsychoSocial Medicine, 4*, 18. https://doi.org/10.1186/1751-0759-4-18

Jover Ibarra, J. (2006). Salud pública y servicios de salud pública. In J. Frutos García & M. A. Royo (Eds.), *Salud pública y epidemiología*. Díaz de Santos.

Kramp, L., & Loosen, W. (2018). The transformation of journalism: From changing newsroom cultures to a new communicative orientation? In A. Hepp, A. Breiter, & U. Hasebrink (Eds.), *Communicative figurations. Transforming communications-studies in cross-media research* (pp. 205–239). Palgrave Macmillan.

López-García, X., Costa-Sánchez, C., & Vizoso, Á. (2021). Journalistic fact-checking of information in pandemic: Stakeholders, hoaxes, and strategies to fight disinformation during the COVID-19 crisis in Spain. *International Journal of Environmental Research and Public Health*, *18*(3), 1227.

Lopezosa, C., Codina, L., & Gonzalo-Penela, C. (2020). *SEO y periodismo: Marco de optimización global como parte del emprendimiento en cibermedios*. Universitat Pompeu Fabra.

Naidoo, J., & Wills, J. (1998). *Practising health promotion: Dilemmas and challenges*. Baillière Tindall.

Newman, N. (2019). Executive summary and key findings. In N. Newman, R. Fletcher, A. Kalogeropoulos, & R. K. Nielsen (Eds.), *Institute. Digital news report 2019*. Reuters Institute for the Study of Journalism.

Observatorio Nacional de las Telecomunicaciones y de la Sociedad de la Información. (2016). *Los ciudadanos ante la e-sanidad*. www.ontsi.es/es/publicaciones/Los-ciudadanos-ante-la-e-Sanidad-0

Pavlik, J. V. (2001). *Journalism and new media*. Columbia University Press.

Ratzan, S. C. (2002). Public health at risk: Media and political malpractice. *Journal of Health Communication*, *7*(2), 83–85. https://doi.org/10.1080/10810730290087969

Revuelta-De-la-Poza, G. (2019). Journalists' vision of the evolution of the (metaphorical) ecosystem of communication on health and biomedicine. *El profesional de la información*, *28*(3).

Rietveld, J., & Schilling, M. A. (2021). Platform competition: A systematic and interdisciplinary review of the literature. *Journal of Management*, *47*(6), 1528–1563. https://doi.org/10.1177/0149206320969791

Rodríguez-González, A. M. (2021). Educación para la salud, prevención y promoción comunitaria a través de la página de Facebook de un centro de salud de atención primaria. *Revista Española de Comunicación em Salud*, *12*(1), 58–66. https://doi.org/10.20318/recs.2021.5307

Ruão, T., Gomes, S., & Silva, S. (2020). Comunicação em Saúde: A cobertura mediática e a gestão da crise COVID-19 numa universidade. *Revista De La Asociación Española De Investigación De La Comunicación*, *7*(14), 54–77. https://doi.org/10.24137/raeic.7.14.3

Salaverría, R. (2019). Periodismo digital: 25 años de investigación. Artículo de revisión. *Profesional De La Información*, *28*(1). https://revista.profesionaldelainformacion.com/index.php/EPI/article/view/69729

Salaverría, R. (2020). Exploring digital native news media. *Media and Communication*, *8*(2). https://doi.org/10.17645/mac.v8i2.3044

Salaverría, R. (2021). Veinticinco años de evolución del ecosistema periodístico digital en España. In R. Salaverría & M. D. P. Martínez-Costa (Coord.), *Medios nativos digitales en España. Caracterización y tendencias*. Comunicación Social Ediciones y Publicaciones.

Sánchez-Bocanegra, C., & Sánchez-Laguna, F. (2012). Las app sanitarias. In I. Basagoiti (Ed.), *Alfabetización en salud. De la información a la acción* (pp. 263–276). ITACA/TSB.

Sánchez-García, P., & Amoedo-Casais, A. (2021). Medios nativos digitales generalistas y especializados. In R. Salaverría & M. D. P. Martínez-Costa (Coord.), *Medios nativos digitales en España. Caracterización y tendencias*. Comunicación Social Ediciones y Publicaciones.

Schiavo, R. (2007). *Health communication from theory to practice*. John Wiley & Sons, Inc.

Singer, J. B., Hermida, A., Domingo, D., Heinonen, A., Paulussen, S., Quandt, Z., & Vujnovic, M. (2011). Participatory journalism. In *Guarding open gates at online newspapers*. Wiley-Blackwell.

Steensen, S., Grøndahl Larsen, A. M., Benestad-Hågvar, Y., & Kjos-Fonn, B. (2019). What does digital journalism studies look like? *Digital Journalism*, *7*(3), 320–342. https://doi.org/10.1080/21670811.2019.1581071

Tejedor, S., Ventín, A., Cervi, L., Pulido, C., & Tusa, F. (2020). Native media and business models: Comparative study of 14 successful experiences in Latin América. *Media and Communication*, *8*(2). https://doi.org/10.17645/mac.v8i2.2712

Van Der Haak, B., Parks, M., & Castells, M. (2012). The future of journalism: Networked journalism. *International Journal of Communications*, *6*, 2923–2938. http://ijoc.org/index.php/ijoc/article/view/1750

Van Dick, J., Poell, T., & de Waal, M. (2018). *The platform society*. Oxford University Press.

Vos, S. C., y Buckner, M. M. (2016). Social media messages in an emerging health crisis: Tweeting bird Flu. *Journal of Health Communication*, *31*, 301–308. https://doi.org/10.1080/10810730.2015.1064495

Westlund, O., Krumsvik, A. H., & Lewis, S. C. (2021). Competition, change, and coordination and collaboration: Tracing news executives' perceptions about participation in media innovation. *Journalism Studies*, *22*(1), 1–21. https://doi.org/10.1080/1461670X.2020.1835526

Zelizer, B. (2017). *What journalism could be*. Polity Press.

Zelizer, B. (2019). Why journalism is about more than digital technology. *Digital Journali Chadwick sm*, *7*(3), 343–350. https://doi.org/10.1080/21670811.2019.1571932

5 Opportunities for information visualization in risk communication

Ángel Vizoso, Gabriela Coronel-Salas and Carlos Toural-Bran

Introduction

Technological development has gradually introduced changes in practically all areas of communication, especially since the beginning of the 21st century (Pavlik, 2001). Thus, some everyday tasks have undergone an innovation that could be described as drastic, while in others, adaptation has been progressive. Of all the revolutions that have taken place over the last thirty years, it is undoubtedly the development of the Internet and all the functionalities and tools associated with it—such as smartphones and social media—that have had the greatest influence on the communication landscape, for purposes both lawful and more harmful to society (Botha & Pieterse, 2020; Freelon et al., 2022; Nguyen & Catalán-Matamoros, 2022).

This influence takes place at all levels, from interpersonal communication to mass communication, whatever its geographical, temporal or thematic scope. In this way, it is possible to appreciate how different areas try to make the most of the full potential of these spaces, both for journalistic and non-journalistic purposes. The Internet has become an essential tool, both internally for organizations and externally in the relationship with their audiences. Something similar is happening with mobile devices. The smartphone is not only an indispensable tool in our daily lives but also humans' most frequent travel companion (Lamsfus et al., 2013), something similar to what happens with social networks. These types of tools open the door to new narratives that seek to exploit the full potential of devices in new environments.

This chapter explores the main features of the use of information visualization as a tool for communication, both in general and more specifically when conveying information to the audience about health or safety risks, among others. Through a review of the potential of data as a raw material for communication, this chapter shows what makes this form of visual communication different from others. For that purpose, Nathan Shedroff's Continuum of Understanding (Shedroff, 1999), a process in which a transfer is executed from raw information elements to an ideal of wisdom or capacity for action and decision on the

DOI: 10.4324/9781032618180-7

content, is set as a basis. The authors try to display not only remarkable projects that are using information visualization as part of their narratives but also some of its current and potential features for the communication of risk.

The power of dataviz

Big data is traditionally defined as a set of "data clusters" of such a voluminous and complex size that they cannot be processed by traditional standard analysis *software*. The challenges facing the computation of this massive data are associated with the capture of the information, its storage, analysis, search and query, transfer, visualization and sharing, updating, information privacy and the sources of origin of the information, among others (Ortega Mohedano & Coronel-Salas, 2019, p. 825).

From data to wisdom

Figure 5.1 depicts a pyramid illustrating the progression from data to wisdom. At the base are data, signifying discontinuous elements representing facts. Moving up, we find information, defined as data processed for practical use. Thirdly, there is knowledge, denoting the mental application of both data and information. Finally, at the pyramid's apex, is knowledge evaluation and internalization, enabling decision-making and action.

To reach the wisdom stage, companies must seize the opportunities of a big data strategy, invest in technologies and training, leverage the investment and

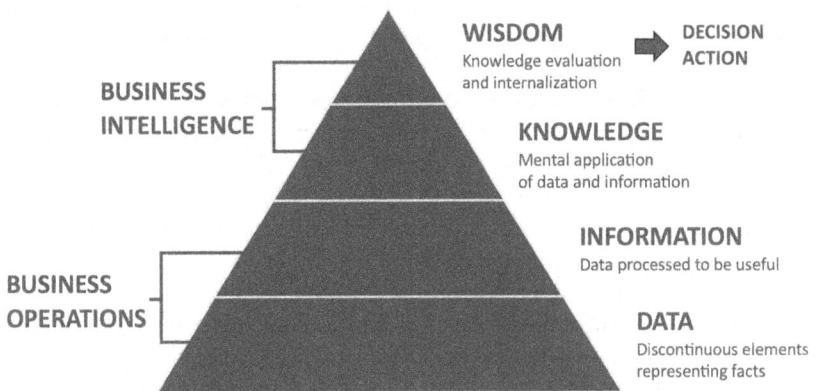

Figure 5.1 "Data, Information, Knowledge, Wisdom (DIKW)" information pyramid.
Source: Authors' own elaboration.

use this "new wisdom" (gained through the process of collecting, cleansing and visualizing data) to improve processes such as customer service and products.

Data are valuable because access to them has the potential to clarify issues in a way that generates results. However, mishandling data can place facts in an opaque structure that communicates nothing. Data may be of limited value to the public if it does not promote discussion or provide an understanding in context.

Display

Data visualization (*dataviz*) translates large data sets and metrics into charts, graphs, pictograms and other visual elements. Data visualization results facilitate identifying and sharing real-time trends, outliers and new insights into the information represented in the data. As the volume of big data increases, more people use visualization tools to access information on their computers and mobile devices. Businesspeople, data analysts, data scientists, data communicators and data journalists use it to make decisions and deliver visually digestible and understandable information to the user. Publishing an Excel table with hundreds of cells is different from publishing an infographic contextualizing the information in that file.

Considering these circumstances, it is possible to highlight the importance of data visualization thanks to elements like the following ones:

- It allows decisions to be made and acted upon quickly.
- It facilitates seeing a complete picture, uncover insights and notice patterns within complex data without relying on a data scientist.
- Presents meaningful data and shares knowledge with others in a way that is easy to understand.
- Democratizes data by providing a source of truth and transparency to users.
- When it comes to knowing the audiences' segmenting audiences/audiences, it is necessary to "draw" their profile through data visualization.
- Contributes to and helps visualize trends in data sets.
- It helps to absorb information faster at different levels.
- It uses the visual language to which society (users) is accustomed.
- Patterns and relationships are identified more quickly using visualizations.

Chart categories

Graphics are divided into categories according to their objectives, aesthetics or visual characteristics (IBM, 2021) and can be versatile and used in different ways, such as:

- Trends
- Comparisons

- Part of a whole
- Correlations
- Relationships and connections
- Maps

Display catalogue

Among the best-known catalogues are:

- The Data Visualization Catalogue: https://datavizcatalogue.com/
- Depict Data Studio: https://depictdatastudio.com/charts/
- Data Viz Project: https://datavizproject.com/

Tools

Thanks to the social web, these types of tools are entirely intuitive, where what is needed is to standardize the database and couple them to the various structures. For this reason, they can be adapted, initially to the work of data visualization, but will become more technical as the effective management of big data progresses. Some of the main tools are:

- Datawrapper: https://app.datawrapper.de/
- Onodo: https://onodo.org/
- Infogram: https://infogram.com/
- Flourish: https://app.flourish.studio/
- Tableau Public: https://public.tableau.com/
- Power Bi: https://powerbi.microsoft.com/
- Google Charts: https://developers.google.com/
- Zoho Analytics: www.zoho.com/es-xl/analytics/
- Qlik Sense: www.qlik.com/es-es/products/qlik-sense
- FusionCharts: www.fusioncharts.com/
- Domo www.domo.com/
- Visme: www.visme.co/

Data-driven products

Four products are used to tell stories with a large amount of data: articles, visualizations, datasets and News Apps.

(1) Articles: they are born from a considerable volume of data and allow us to build stories; above all, they are short articles, and most use text detailing figures, values, numbers and percentages, among others.

- La Nación (Jastreblansky, 2012): "Los millones de la APE: cómo se repartió en 2011 la caja que era de Moyano"

- La Nación (Alconada Mon et al., 2021): "Pandora Papers: an offshore adds mystery to the disappearance of US$7 million in the Baggio juice company"

(2) Data visualizations (dataviz): they show information obtained through large volumes of data through images, graphics, infographics, pictograms and other elements. Short or long articles can complement them and, in some cases, be accompanied only by a headline and an explanation so that users can analyze and read them.

- La Nación (2021): "The evolution of the pandemic in Argentina"
- *Primicias*.ec: "More than 159,000 people ceased to have adequate employment in October"

(3) Datasets: as its name suggests, it is a collection of data or a set of data that allows us to access information for processing. In a certain sense, it is usually linked to the transparency of organizations. We can access databases on the Internet according to the subject we are looking for, but this will depend on who provides them and their transparency to the community, which is considered "open data."

- Google Dataset Search: https://datasetsearch.research.google.com/
- Tableau: www.tableau.com/es-es/learn/articles/free-public-data-sets

(4) News Apps: these applications concentrate and focus a story on a single space to tell it far beyond a single article. According to Sandra Crucianelli (2013), "many times, the volume of data is so large that it is impossible to find a news story if an application is not designed to group and analyze variables, for example, by geolocation, by date, by company name, etc."

- The Texas Tribune: https://salaries.texastribune.org/
- ProPublica: https://projects.propublica.org/docdollars/

Work phases

Some media outlets have created innovation laboratories (Labs), including developing new products focused on a hypermedia narrative. As seen in Figure 5.2, it is essential to work together with several professional profiles of the media and, above all, to comply with the basic precepts of journalism: verify and contrast information. Among the work phases to implement a product with data, there are six shown in Figure 5.2.

Telling stories with data

For John Bones of the Verdens Gang, many times, it is not necessary to visualize a story, but the media often do so for readers or users to size up a fact. While stories containing large data often need visualization, we must be critical

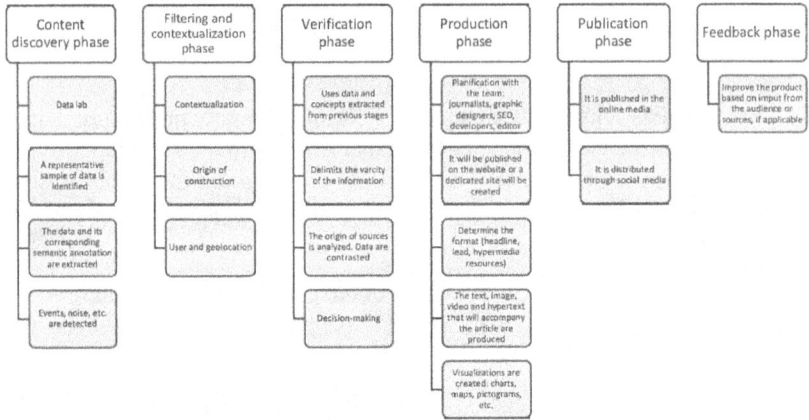

Figure 5.2 Phases of work with data.

Source: Authors'own elaboration based on the ideas of authors specialized in data.

in choosing the data we will present. When we report on something, we know all sorts of things, but what does the reader need to know about the story? A table or a simple graph showing a process going from year A to year C will suffice. When working with data journalism, the goal is to present a manageable amount of data. It is about journalism.

A good visualization is like a good picture. You understand what it is all about just by looking at it for a moment or two. The more you look at the visualization, the more you see. Visualization is better when the reader needs to know where to start or where it ends and when the visualization is overloaded with detail. In this case, a piece of text would be better, as concluded by Bones (Bounegru et al., 2012, pp. 215–216).

Storytelling

Storytelling translates as "the art of storytelling" because it is important to connect emotionally through a story, to let it touch the heart, the mind, the body and the spirit: the rational and the instinctive. A good story creates a positive emotion that inspires people to take action. It is not about showing a brand; it is about telling what experience the user will feel with that product/service. To do this, it is essential to "think like a user" and to apply *storytelling* through the various platforms (including visualization). It is essential to know the audience to whom the communication product will be addressed. To create products based on data, it is essential to think about the final recipient, i.e. the public, the audience, the user: *a) occasional:* they want information simply and quickly. They are interested in having an idea of the data, not a detailed analysis; *b) active:* they stimulate debate and use the data to increase their knowledge of a given area; *c) hoarders:* they want raw data to make their visualizations or analysis.

Steps to create a visualization

As shown in Figure 5.3, the data visualization expert Alberto Cairo (2014) highlights the existence of seven steps in the elaboration of a data visualization project.

As there are no absolutes in life and even less so in digital communication, the proposal of seven stages to carry out a visualization proposed by Ann K. Emery (2014; cited in Grassler, 2017) is shown in Figure 5.4.

It is necessary to point out that, traditionally, a media outlet, through its journalistic team, was introduced in the search for data that would allow it to verify and test its hypotheses. Nowadays, the "traditional" process is maintained, but technology is added through big data, including artificial intelligence (AI), for information analysis, development and evaluation. Production in the journalistic

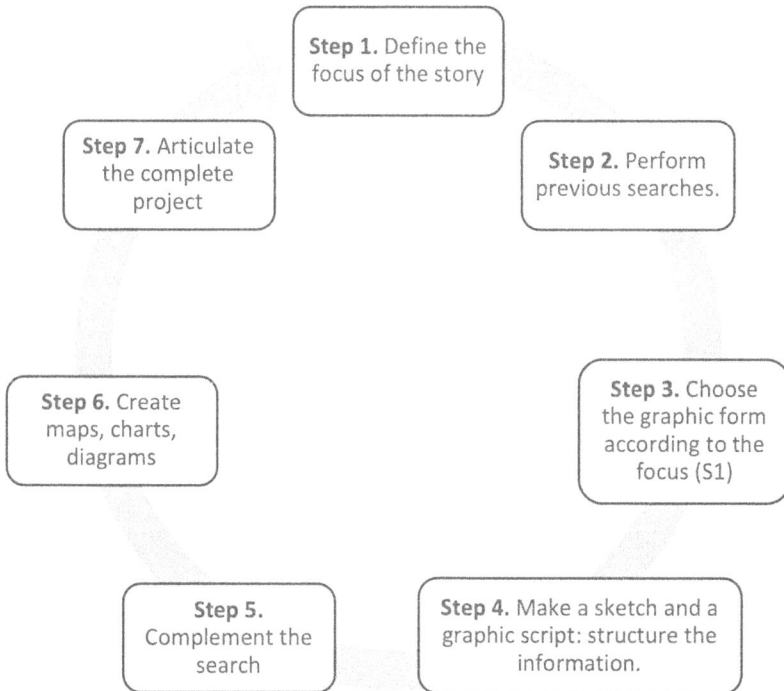

Figure 5.3 Steps to create a visualization.

Source: Authors' own elaboration based on Alberto Cairo (2014; in Grassler, 2017).

Stage 1. Select a single message to define the graph	• Determine the audience • Choose the best graph for the data • Determine the needed level of detail
Stage 2. Reduce clutter	• Delete or lighten the edges, the grid lines, or the marks on the axis. • Examine every ink point on the graph.
Stage 3. Direct labeling	• Add the needed data to help the comprehension in the bars and lines of the graph to avoid repeating information.
Stage 4. Emphasize key findings with colors.	• Choose a customized color palett. • Emphasize the message with color.
Stage 5. Show the main message in the title	• Use 6 to 8 words in the title. • Use 2 or 3 phrases for the title.
Stage 6. Are you doing it right? Try it.	• The strabism test • Test with other contacts/users/colleagues.
Stage 7. Adapt and share the graph	• Charts within the live presentation or webinars • Graphics like the star of a web site • Graphics shared on Twitter • Graphics in papers and infographics

Figure 5.4 The seven stages of visualization.

Source: Authors' own elaboration based on the proposal of Ann Emery (2014; in Grassler, 2017).

field, the use of methods of weighting and collecting information, has evolved: news is emerging from the numbers and not the other way around. Not to mention that thanks to this type of technology, new roles have been created, such as data science, data architect, big data consultant, data analyst and data scientist, among others.

Let us remember that where there is data, there is information; where there is information, there is knowledge and, therefore, wisdom.

Opportunities for information visualization in risk communication

This section reviews some of the elements that need to be considered when assessing the potential of information visualization as an element for risk communication. Hence, it seeks to point out these elements through examples of the application of some of its characteristics in the field of risk communication. Nonetheless, the following lines do not only focus on its potential in the media but also as an instrument for visual communication in all types of organizations.

The role of public sources

Information visualization, especially in its performance in the field of data, has benefited enormously from the expansion of the recording and use of this type of information unit. In fact, some research points out how today's society is immersed in a process of datafication, both on a general level (Arsenault, 2017) and in the field of media (Ausserhofer et al., 2020). This circumstance provides the press with a powerful source of content while, at the same time, forcing it to update routines and narratives in order to share them with the audience.

In this regard, narratives such as the one addressed in this chapter have emerged as flagships of this form of journalism, as they have the capacity to host high doses of content—in the form of data, in this case—while allowing for its exploration. Furthermore, for a little over a decade, a large part of the data that has been entering journalistic information as a source has been in the form of Open Data. In addition to being a magnificent resource for society in general and professionals or researchers in any sector, this type of space is a resource of great value for journalism and, more specifically, for the visualization of information (Veira-González & Cairo, 2020, p. 140).

If we look specifically at risk communication, it also seems reasonable to point to public sources as central elements of this narrative. If, moreover, we do so in terms of visual communication, it is necessary to focus attention on the potential of the use of public sources in information content of this nature.

A good example of the use of public data sources and their combination with information visualization—in addition to other narratives—is the automated tool Quakebot, developed by the Los Angeles Times' Data and Graphics Department. This automated bot navigates the databases of the US Geological Survey, detecting earthquakes. Once identified, a piece of basic information is produced and presented with a map. We are, therefore, faced with the combination of two of the fastest-growing journalistic narratives in recent years. Firstly, automation is an evolving resource that opens up new possibilities for journalistic communication. Secondly, the information visualization which, in this case, is presented accompanying and providing geographical context to the content created by the bot.

A high level of adaptability

The term adaptability is one of those that best fits when talking about the visualization of information as a communicative tool. Thus, it is possible to speak of its capacity to settle in multiple media, from the most traditional—such as print—to those already characteristic of the digital scenario (Dick, 2013; Pérez-Seijo & Vizoso, 2021). Furthermore, if we make this observation from the thematic level, the visual presentation of information serves as an element of transmission for multiple types of information, from the simplest to the most complex. In fact, this ability to hold content with a high level of complexity is one of the main values of the visual treatment of information and one of the reasons why it is often used (Anne DiFrancesco & Young, 2011).

Within the international journalistic panorama, if we look at examples of the production of content aimed at presenting or explaining risks of various kinds to the public, we find the case of Reuters Graphics (*www.reuters.com/graphics/*). This journalistic communication giant uses a large part of this fully visual space to produce content aimed at dealing in depth with certain problems with global repercussions.

One good example is the project "The Hottest Year," published in December 2023 and authored by Gloria Dickie, Travis Hartman and Clare Trainor. In this piece, a classic element of journalism, such as the word, does not disappear, although in this case, it is not the main character given the high relevance of visual elements—basically charts and maps—in the construction of the story. The high information load present through data relating to temperature or the location of adverse meteorological phenomena such as hurricanes or tropical storms makes it necessary to use efficient and easily understandable formulas for the audience. In this sense, the visualization of information manages to provide this efficiency, but with a very large amount of content inside.

Also, in the field of adaptability, it is necessary to point out the capacity of this form of communication and journalism to be presented in almost any medium. In this respect, the visualization of information follows the line marked in the production of many other news contents, in which the ideas of mobile journalism, the change in the consumption habits of audiences (Ghersetti & Westlund, 2018) and the work for multiple devices mark the way forward. In relation to this idea, it should be noted that interactivity within the proposed charts or maps is not usually the main feature. Generally, the bet revolves around projects in which there is a high load of graphics—which usually exceeds the textual content—but in which either a single layer is presented or, in any case, a succession of layers in the form of animation or through the creation of a GIF. Notwithstanding the fact that the media and its professionals have the skills to produce fully interactive discourses within their graphics, these are sometimes sacrificed for ease of consultation on smaller screens such as mobile phones, where navigation is not as easy as on the computer.

Visual risk communication outside the media

Although, as detailed previously, the use of visual narratives for risk communication is one of the usual practices in this type of communication within journalism, this type of presentation is not exclusive to the media. Especially in recent years, as a consequence of the COVID-19 pandemic, many institutions and public bodies have made use of the visualization of information when alerting or communicating certain risks to the population. A clear example is, as detailed previously, the development of dashboards by organizations such as the World Health Organization (WHO) (*https://covid19.who.int/*). This space, constructed almost entirely through the use of graphs, tables and maps, is intended for the detailed presentation of a large volume of information.

The global nature of the COVID-19 outbreak necessitated the development of solutions of global scope and application. Spaces like this one, present in organizations such as the Johns Hopkins University & Medicine in the United States (*https://coronavirus.jhu.edu/data*) or the Instituto de Salud Carlos III in Spain (*https://cnecovid.isciii.es/covid19/*), among others, speak of the potential of visual presentation of data in a time of health crisis. This type of approach benefits not only from the possibilities of exploring figures and the interactive capabilities that can be applied to the content created but also from the efficiency of visual content in capturing the audience's attention and generating recall (Borkin et al., 2015).

Final considerations

As highlighted, data visualization has great potential in the field of journalism. It is a current reality when it comes to approaching and producing news content so the media have been integrating it into their narratives. The visual presentation of information connects directly with the process of increasing the recording and availability of data, simplifying its presentation not only to media audiences but to society in general.

If we look exclusively at risk communication—health, environmental, etc.— all this potential, which is valid for any type of information, can take on a new dimension. In the scenario of communicating this type of content, which is so complex and often far from people's knowledge, it is vital to opt for genres that are efficient when it comes to communicating but also attractive when it comes to generating audience recall. Thus, the combination of the characteristics of the visualization of information, the availability and ease of access to tools that make its creation possible and the commitment to these narratives in the media leads to the identification of a great potential as an instrument for journalistic content that not only seeks to inform, but also to alert and raise awareness of risk in its various presentations.

Moreover, this form of visual communication of information has leapt beyond the journalistic scene, gaining great importance as an instrument for the dissemination of data by public organizations. In this sense, as described in this chapter, the COVID-19 pandemic has been the most recent and most notable example. In short, the visual presentation of information serves as a tool to facilitate access to complex information by society. While this feature is relevant in any subject matter, it is even more so when the aim is to communicate useful and relevant information in order to limit or minimize the effects in circumstances of risk.

Acknowledgments

This chapter is part of the project *Radon in Spain: Public perception, media agenda and risk communication* (RAPAC), financed by the Spanish Nuclear Safety Council [Consejo de Seguridad Nuclear] (SUBV-13/2021).

References

Alconada Mon, H., Jastreblansky, M., & Brom, R. (2021, November 1). Pandora Papers: Una offshore suma misterio a la desaparición de US$7 millones en la empresa de jugos Baggio. *La Nación.* https://www.lanacion.com.ar/politica/pandora-papers-una-offshore-suma-misterio-a-la-desaparicion-de-los-us7-millones-en-la-empresa-de-nid01112021/

Anne DiFrancesco, D., & Young, N. (2011). Seeing climate change: The visual construction of global warming in Canadian national print media. *Cultural Geographies, 18*(4), 517–536. https://doi.org/10.1177/1474474010382072

Arsenault, A. H. (2017). The datafication of media: Big data and the media industries. *International Journal of Media and Cultural Politics, 13*(1–2), 7–23. https://doi.org/10.1386/MACP.13.1-2.7_1

Ausserhofer, J., Gutounig, R., Oppermann, M., Matiasek, S., & Goldgruber, E. (2020). The datafication of data journalism scholarship: Focal points, methods, and research propositions for the investigation of data-intensive newswork. *Journalism, 21*(7), 950–973. https://doi.org/10.1177/1464884917700667

Borkin, M. A., Bylinskii, Z., Wook Kim, N., May Bainbridge, C., Yeh, C. S., Borkin, D., Pfister, H., Member, S., & Oliva, A. (2015). *Beyond memorability: Visualization recognition and recall.* https://vcg.seas.harvard.edu/files/pfister/files/infovis_submission251-camera.pdf

Botha, J., & Pieterse, H. (2020). Fake news and deepfakes: A dangerous threat for 21st century information security. *Proceedings of the 15th International Conference on Cyber Warfare and Security, ICCWS 2020,* 57–66. https://doi.org/10.34190/ICCWS.20.085

Bounegru, L., Chambers, L., & Gray, J. (Eds.). (2012). *The data journalism handbook 1.* European Journalism Centre. https://datajournalism.com/read/handbook/one

Crucianelli, S. (2013). ¿Qué es el periodismo de datos? *Cuadernos de Periodistas, 26.* www.cuadernosdeperiodistas.com/que-es-el-periodismo-de-datos/

Dick, M. (2013). Interactive infographics and news values. *Digital Journalism, 2*(4), 490–506. https://doi.org/10.1080/21670811.2013.841368

Freelon, D., Bossetta, M., Wells, C., Lukito, J., Xia, Y., & Adams, K. (2022). Black trolls matter: Racial and ideological asymmetries in social media disinformation. *Social Science Computer Review, 40*(3), 560–578. https://doi.org/10.1177/0894439320914853

Ghersetti, M., & Westlund, O. (2018). Habits and generational media use. *Journalism Studies, 19*(7), 1039–1058. https://doi.org/10.1080/1461670X.2016.1254061

Grassler, M. (2017). *El rol del periodista de datos en el proceso de los sistemas de gestión y decisión pública y en la recuperación de la confianza entre el ciudadano y las instituciones públicas.* Universitat Autònoma de Barcelona.

IBM. (2021). *What is data visualization?* www.ibm.com/topics/data-visualization

Jastreblansky, M. (2012, June 25). Los millones de la APE: Cómo se repartió en 2011 la caja que era de Moyano. *La Nación*. www.lanacion.com.ar/politica/los-millones-de-la-ape-com o-se-repartio-en-2011-la-caja-que-era-de-moyano-nid1484852/

La Nación (2021, November 25). La evolución de la pandemia en la Argentina. *La Nación*. www.lanacion.com.ar/sociedad/en-detalle-infectados-fallecidos-coronavirus-argentina-nid2350330/#/

Lamsfus, C., Xiang, Z., Alzua-Sorzabal, A., & Martín, D. (2013). Conceptualizing context in an intelligent mobile environment in travel and tourism. In L. Cantoni & Z. Xiang (Eds.), *Information and communication technologies in tourism 2013* (pp. 1–11). Springer. https://doi.org/10.1007/978-3-642-36309-2_1

Nguyen, A., & Catalán-Matamoros, D. (2022). Anti-vaccine discourse on social media: An exploratory audit of negative tweets about vaccines and their posters. *Vaccines*, *10*(12), 2067. https://doi.org/10.3390/VACCINES10122067

Ortega Mohedano, F., & Coronel-Salas, G. (2019). Big data, augmented data y computación cognitiva en la era del millenial. In L. M. Romero Rodríguez & D. E. Rivera Rogel (Eds.), *La comunicación en el escenario digital: Actualidad, retos y prospectivas* (pp. 821–853). Pearson Educación de Perú.

Pavlik, J. V. (2001). *Journalism and new media*. Columbia University Press.

Pérez-Seijo, S., & Vizoso, Á. (2021). ¿Infografías en los reportajes en vídeo 360º? La integración de la visualización de la información en entornos esféricos. *Estudios Sobre El Mensaje Periodístico*, *27*(2), 607–622. https://doi.org/10.5209/ESMP.70547

Shedroff, N. (1999). Information interaction design: A unified field theory of design. In R. E. Jacobson (Ed.), *Information design* (pp. 267–292). The MIT Press.

Veira-González, X., & Cairo, A. (2020). From artisans to engineers. How technology transformed formats, workflows, teams and the craft of infographics and data visualization in the news. In C. Toural-Bran, Á. Vizoso, S. Pérez-Seijo, M. Rodríguez-Castro, & M. -C. Negreira-Rey (Eds.), *Information visualization in the era of innovative journalism* (pp. 134–153). Routledge.

6 Health communication, awareness raising, and metaverse

An approach to the radon gas situation in Spain

Pavel Sidorenko Bautista,
Jessica Zorogastua Camacho
and Mariola Moreno Calvo

Introduction

In recent years, the world has faced major health emergencies at both national and international levels, epitomized by the COVID-19 pandemic. In many cases, government preparedness to deal with these emergencies has been inadequate. However, effective government communication is critical to educating the public about impending threats and best practices for mitigating damage during such emergencies (Do Kyun & Kreps, 2020). A crucial facet of risk communication is understanding how the framing of risk-related messages affects public perception (Freudenstein et al., 2020).

Many experts argue that risk acceptability depends on two main components: danger and outrage. The number of exposed, infected, and sick people denotes danger. Conversely, public and patient reactions to risk mitigation messages correlate with outrage (Malecki et al., 2021), which derives the need for efficient risk communication in the face of extreme and sudden danger, such as the outbreak of a deadly disease or any other phenomenon that affects people's stability or lives (Heydari et al., 2021).

Currently, social networks provide a platform for experts to convey hazard information and provide beneficial advice quickly. However, these platforms are also breeding grounds for misinformation, underscoring the need to monitor content to avoid unwarranted panic and erroneous practices detrimental to public health (Poland & Spier, 2010; Swire & Ecker, 2018; in Lu, 2020). It also calls for exploring alternative digital avenues to counteract the effect of distortion and lying. Nevertheless, the ideal is—if possible—to prevent scenarios through communication focused on awareness, prevention, and education, especially in the face of persistent problems such as continued exposure to a harmful gas, as in the case of radon.

The following lines offer a vision of the problem specifically in Spain and a proposal focused on orienting and educating emerging audiences through disruptive alternatives such as the metaverse.

DOI: 10.4324/9781032618180-8

Implications of radon gas for human health and the situation in Spain

In 2021, the World Health Organization identified radon gas as one of the leading causes of lung cancer. It estimated between 3% and 14% of the proportion of lung cancer cases that could be attributed to this gas, depending on their average concentration and the prevalence of tobacco use in each country. Different studies show that a smoker is 25 times more likely to have cancer if exposed to radon and that the risk of this type of cancer increases in the population by 16% with each increase of 100 Bq/m3 in the average concentration of this gas in the long term; thus there is a linear relationship: the risk of cancer increases in proportion to increased exposure to gas (WHO, 2021).

Radon gas, which is produced by the natural radioactive disintegration of uranium, may be present in soil and water. However, there are no epidemiological studies that show that its presence in drinking water is related to some type of cancer. On the contrary, their presence in the soil affects humans more directly. According to the *WHO Handbook on Indoor Radon: A Public Health Perspective* (WHO, 2009), the greatest exposure to radon occurs in homes and indoor workplaces, where many hours of daily life are often spent.

Depending on the characteristics of the soil where this gas is located, it can be filtered into homes and buildings by cracks, pipes, drains, pores of constructed walls, etc. Their concentration can vary between buildings, within the same house, and from one day to the next; therefore, gas measurement is promoted in indoor spaces with passive detectors.

Radon in Spain: geographic location and main health effects

Spain has 17% of its national territory affected by radon gas, where its potential exceeds 300 Bq/m3, which is the reference level established by Directive 2013/59/ Euratom for radon concentration to be considered a priority action area. By surface area affected, the regions with the highest radon concentration areas are Galicia, with 70%, Extremadura, with 47%, Madrid has 36%, and Castilla y León and the Canary Islands, which have 19% each. The rest of the regions present values close to or below 10% (Ministerio de Sanidad, 2021). Due to the high presence of Radon in Galicia, the University of Santiago de Compostela has promoted a specific laboratory on the subject, which, through the performance of thousands of measurements, has allowed it to elaborate its own Galician Radon Map.

Although Spain as a member country of the European Union is subject to the Community guidelines on Atomic Energy and risks from exposure to ionizing radiation (2013/59/Euratom), its action is partial according to Royal Decree 1029/2022 (BOE, 2022, December 20), which results in the absence of a National Plan and a technical building code, mandatory according to the 2013 Directive, although experts point out that protection against radon was already implemented in homes after that date (Fernández, 2022).

The Spanish Ministry of Health considers it important to raise public awareness on this issue, even though

There is no evidence that any study has been conducted in Spain to measure the degree of awareness or sensitivity of the population regarding the risks associated with radon exposure and prevention measures that would allow the design of more specific strategies.

(Ministerio de Sanidad, 2021, p. 32)

On this background lies the importance of having information and awareness campaigns, especially in areas with a high concentration of this gas, such as Galicia, Castilla y León, Extremadura and the Canary Islands, where more than 10% of the buildings may present, on the ground or second floor, Radon concentrations above the reference level—according to the potential radon map in Spain (*https://bit.ly/3PZUbCO*) prepared by the Spanish Nuclear Safety Council (CSN) in collaboration with different universities. Perhaps the greatest limitation in this respect refers to the fact that these are regions with a large rural area, lower population density, and/or older population (INE, 2023).

In addition to being the most affected by radon gas, these regions make the least use of the Internet. According to the latest Survey on Equipment and Use of Information and Communication Technologies (ICT) in Spanish households, which indicates that the use of ICT by people aged 16 to 74 years reaches 94.5%, that is, 33.5 million people, these three regions are about three percentage points below the average and up to eight concerning the region that makes more use of this technology (INE, 2022), and which highlights that from 55 years individuals access the Internet less frequently.

In Europe, approximately 20% of all cancer deaths are caused by lung cancer (Baum et al., 2022), with smoking being the most important risk factor. However, 15% to 25% of all lung cancer cases are diagnosed in people who have never smoked (Torres-Durán et al., 2014).

Thus, radon gas is the second cause of lung cancer after smoking and the first in non-smokers (Barros-Dios et al., 2012). According to the World Health Organization, Radon is estimated to cause between 3% and 14% of all lung cancers in a country, depending on the national average level of radon and the prevalence of smoking (WHO, 2009) and in Spain, it accounts for 3.8% of all lung cancer deaths.

The risk of lung cancer increases statistically significantly in proportion to Radon exposure, although other pathologies are still unknown (Ngoc et al., 2023). A study of miners found that those who had been exposed to Radon 5 to 14 years previously and miners younger than 55 years had the highest risk of lung cancer mortality (Lane et al., 2019). Children represent another of the most sensitive segments to this scenario because they have a higher number of dividing cells compared to adults (Ruano-Ravina et al., 2018).

Curiously, even though it represents a relevant problem in terms of public health, not only are there few concrete preventive actions regarding radon gas in Spain, but media coverage does not accompany this situation either (Negreira-Rey & Vázquez-Herrero, 2022).

Awareness and communication in the digital and virtual environment: actors and current cases in extended reality

As in other professional fields, extended reality (XR) (the set of contents and experiences developed in virtual reality—VR, augmented reality—AR, and/or mixed reality—MR) has found an important functionality in the health sector, through training proposals, treatments (surgery, psychiatry, etc.), or simply of an informative nature.

For example, residents at the Cleveland Clinic benefit from MR through Microsoft HoloLens viewers to receive better training in anatomy (see *https://youtu.be/h4M6BTYRlKQ*). For its part, the AccuVein mobile app uses RA to enable professionals in training or exceptional cases to distinguish the best place to proceed with a blood sample collection (see *https://youtu.be/SnSj11oJV-0*). For its part, the San Juan de Dios Residence for the Elderly, in the city of Granada (Spain), carries out a program for the treatment of senile dementia and other psychiatric pathologies through virtual experiences using immersive viewers (see *https://youtu.be/rKS7uhfRPRo*).

There are multiple applications in three-dimensional surgical care or palliative experiences with renal or oncology patients. However, in the specific realm of health communication, particularly in the field of awareness, what are the specific uses? As virtual reality is specifically a technology that resorts to immersion with which to try to bring the user closer to the content, the chances of developing greater empathy on the part of the audience are increased (Serino & Repetto, 2018; Herranz & Bautista, 2023; Ventura et al., 2019). That said, from the field of NGOs (García-Orosa & Pérez-Seijo, 2020), the media and institutions of various kinds have been working on storytelling to raise awareness through 360° immersion and virtuality on several problems of a social nature (see *https://youtu.be/dAqK9lXdVxM*). Moreover, the Carlos Slim Foundation's Digital Health program (see *https://bit.ly/45WeoiH*) specifies and is convinced that the metaverse will represent a solution in terms of health communication and guidance and awareness initiatives.

Difference between virtual reality and metaverse

Once the concept of extended reality, specifically virtual reality, has been defined, it is essential to define what the metaverse consists of and how it intervenes in this digital ecosystem with immersive possibilities. Unlike a "traditional" virtual world, the metaverse offers open platforms that allow users to develop and share content and applications. It also offers these users the opportunity to carry out commercial operations and other rewards for creating and sharing content, which is not typical of classic virtual worlds (Moioli, 2022).

Metaverse comes from a composition of two words: "meta", meaning "beyond", and "universe", alluding to a virtual world where users interact and perform activities as they wish (Changhee, 2021). It could be summarized as a digital extension of any social, productive, or entertainment action carried out by human beings, i.e., it is a virtual world in which people are represented by cyber identities (Abbott, 2007). Real-time interaction is one of the main variables of this virtual instance.

The idea of community and the possibility of socialization are closely related to the metaverse, though they are not defining characteristics of virtual reality experiences; quite the opposite. According to Moioli:

> Another difference between the worlds of the metaverse and virtual reality is a phenomenon known as the Internet of Things. This is the integration of the physical world with the virtual world using objects equipped with sensors and connected to the Internet. When we talk about augmented reality or mixed reality, the physical and the digital are seen and experienced simultaneously. If I have a meeting where I see some colleagues physically present in the room with me and others sitting next to me as avatars, the very question of "How relevant in the future might our online life be compared to our physical life" totally loses its meaning. This is the metaverse.
>
> (Moioli, 2022)

The author believes that the metaverse provides an unprecedented level of virtual interaction and collaboration and makes it a "highly customizable" space. With this, it is possible to affirm that users have the availability to modify and improve the metaverse itself, as well as their avatars.

In contrast, conventional virtual reality experiences simply ensure that users can explore and interact with the available environments, but always according to the developers' and manufacturers' wishes and criteria. According to the postulates of Castronova (2001), for a virtual experience to be considered a metaverse, it must meet three basic conditions: interactivity in real-time (with other users and with the environment), corporeality (avatars), and persistence (continuous connection without interruptions).

Accenture (2022) prepared a report on the emergence and development of the metaverse. This document states that it is part of the so-called "Extended Reality"; it is increasingly trying to combine real experiences with the virtual ("virt-real") and is a determining factor in the evolution of some business models.

For all these reasons, they have coined the concept of "continuous metaverse", i.e., a metaverse in constant evolution.

Towards a communication model on radon gas in the metaverse

The metaverse is nothing more than a medium through which various narratives and experiences can be developed to reach the desired audience. As already stated, it is nothing more than a conglomerate of virtual environments with their characteristics of immersion, dialogue, and interaction. As with social networks and digital platforms, in the metaverse, there will be the possibility of having open and public options, but it also allows the possibility of providing closed and controlled spaces for specific purposes if the characteristics already listed in previous sections are respected.

Target audiences for the metaverse communication

Perhaps the main—but not the only—antecedents of the metaverse refer to "Second Life" and Roblox at the beginning of the 21st century. Both platforms were a reference for Millennials, i.e., users who experienced the digital migration to social networks from 2007 onwards; users with a strong and important digital culture (Lu, 2021). Currently, this audience is associated with the acquisition and creation of digital goods for the home, as well as access to virtual events and cryptocurrency transactions (YPulse, 2022).

But this context harbours a complexity in this regard, and it is that other segments cohabit and interact virtually, as is the case of centennials and alphas: the former also called "digital natives" (Baysal, 2014; Prensky, 2001; Turner, 2015), and the latter, children and adolescents defined by streaming and digital-mobile interaction (McCrindle & Fell, 2021). This context is also determined by video games—or platforms—such as Fortnite or Minecraft, which have provided users with the possibility of acquiring and exchanging virtual goods and accessories as a habitual and even necessary activity.

According to JP Morgan (2022), at the beginning of the third decade of the 21st century, investments of more than 54 billion dollars have been reported around the metaverse, which allows us to think that we are not only in an evolution of virtuality but of the immediate future of the Internet as we know it today. The "Alpha" generation (born after 2010) is investing more and more time and money in virtual activities and platforms, which makes them its main target audience (Sidorenko & Herranz, 2021).

Any action to be developed in the metaverse must, of course, take this into account to procure experiences that address these profiles. Evolution continues, and communication strategies must adapt as quickly as possible in the virtual environment as in the real one. In other words, communication actions focused on awareness-raising must consider this heterogeneity of segments, while efforts aimed at staff training or coaching must assess the profile of this target audience.

Between awareness and professional training: references and examples

At present, most open metaverses are designed for playful experiences or focused on new dimensions of socialization and dialogue between users, which is an important premise and variable for awareness-raising discourse. Current theories on learning from virtual experiences explain how users can become more engaged with content and thus learn through reflection, verbal interactions, mental operations (e.g., decision-making), and vicarious experiences, such as a virtual hospital in which learners make certain decisions based on contexts and cases (Loke, 2015).

However, contrary to what Drake et al. (2011) or Vergara et al. (2008) claim, a virtual experience will never be a substitute for a real one but will represent a reinforcement or enrichment experience about a certain knowledge. For Loke (2015), it is worth asking whether "learning by doing" (Dede, 1995) is the best way to procure a learning process through semi-immersive virtual experiences (via PC, mobile devices, or game consoles), given that the main interaction would be verbal because the avatar's bodily actions would be subordinated to a keyboard or controller, instead of the user's direct involvement.

But in all cases, there is greater user involvement and less chance of procrastination and/or distraction, which leads us to ask, what would be the best way to raise awareness of radon gas, through virtual awareness campaigns or virtual learning modules? In this case, it could be approached as a corporate communication campaign in both external and internal terms. On the one hand, there would be the citizens who should be informed and educated on the subject. On the other hand, in the affected communities, the corresponding authorities and related public institutions (municipalities, ministries, etc.) would be another instance. Therefore, the main limitation is technological, either due to a literacy issue (the metaverse concept is not yet properly massified, and the older population is less and less able to assimilate these changes) or due to the speed of connections (the optimal performance of the metaverse depends on a good bandwidth or, failing that, on a 5G mobile connection).

In the industrial field, training metaverses where users interact in real-time with other colleagues and with their trainers have been implemented for some time (Nafarrete, 2017). A Spanish example is the state-owned railroad company Renfe (Godoy, 2023). Although in the purely virtual field, it has been known that for some years, the military sector (Harris et al., 2023), manufacturing (Ulmer et al., 2020), and the retail sector value the virtual relationship as an important resource to recruit human capital and provide it with certain skills before facing real cases. In particular, the healthcare sector has also benefited from virtual and immersive narratives for educational and training purposes, as in the case of renowned hospitals such as Johns Hopkins or Mass General, both in the United States (Horowitz, 2022).

Metaverse and radon gas: model proposals

In the field of health communication, or communication from the health sector, there are already some references specifically related to the metaverse. It should be noted that most of these cases are strictly related to the hospital sector, specifically to the offer of services or to reputational communication.

By way of example, to mention a few cases, in Spain, the San Juan de Dios Hospital in Zaragoza was among the first to develop a digital twin through which it has various institutional information, services, schedules, and even a transparency portal (see *https://virtual.hsjdzaragoza.es*). Even on the Internet, it is possible to come across the "Metaverse Hospital" (see *www.hospital-metaverso.com*), which, although it does not specify how the virtual interface works, does require secure ways of establishing contact with patients, as well as operating with monetary operations through crypto-assets and virtual wallets. Another reference to healthcare services is in the United Arab Emirates (EHS). The metaverse is enabling them to implement telemedicine through virtual resources while trying to "get closer to the patient" (see *www.ehs.gov.ae/en/metaverse*).

Alex Jakma (2022) points out that the metaverse allows for reaching global audiences, creating disruptive experiences, and, therefore, generating interest in the brand, as well as trying to improve the relationship with customers through avatars, content strategy, digital assets, and community management. These actions can be carried out through posters in the form of images embedded in virtual universes, the design of virtual wearables, gamification, and immersive experiences.

Metaverse environments such as Fortnite or Roblox, with high penetration worldwide, can be used to promote informative campaigns because of the impact they have on family or educational dynamics (Victoria González, 2020; Bonelo & Amar, 2023). The persuasion in Fortnite has already been widely studied, from the actions of commercial brands to political communication (Soto de la Cruz et al., 2023).

In the specific case of radon gas, there is a virtual reality reference—not in the metaverse as such—developed by the French General Directorate for the Environment, Spatial Planning and Housing, in which users must solve a series of puzzles about this problem to escape from a room.

In short, today, in the metaverse, there are actions from professional fields of various kinds, but there is no evidence around public health and scientific awareness. In other words, there is still no evidence of operational "scientific health" metaverses. Nor is there evidence of awareness campaigns designed expressly for the metaverse on these issues.

The metaverse as a channel for radon gas awareness in Spain

Awareness campaigns in the metaverse should focus on users most accustomed to the virtual proposal and, simultaneously, most active within it. Therefore, efforts should initially target the Alphas. Consequently, two options are available: an informative documentary proposal and a gamified one that engages the user in the proposed problem.

For the first case, the document on indoor radon gas (NTP 440) prepared by the Spanish Ministry of Labour and Social Affairs, in collaboration with the National Institute for Safety and Hygiene at Work, could be used as a reference. The infographic used (refer to Figure 3 at *https://bit.ly/3LKrRC4*) is considered highly useful. It would be advantageous to transfer it into an interactive 3D model in which avatars can move freely. This would enable users not only to view information regarding the routes of gas entry in the building but also to access reinforcement information on actions and prevention. Moreover, it could serve as a disruptive resource for teachers and trainers, providing them with the opportunity to organize an expedition to a restricted access place without compromising the integrity of any of their students.

Regarding the second proposal, about the escape room developed by the French environmental authorities, as mentioned earlier (see *https://bit.ly/3ZFwjaY*), the concept would be to replicate it in the metaverse as a gamified knowledge reinforcement technique. This approach aligns with virtual experiences in Fortnite (makerspace) or Roblox.

Metaverse as a training platform for radon gas actions

A training metaverse is not far from the model of applying virtual reality for such purposes, i.e. it must have a well-defined call to action and solve the problems to develop or reinforce users' competencies in the future in the real world. This said, the virtual must have a positive and transcendental qualitative impact on the real performance of human capital.

This virtual training dynamic must target audiences less connected to the metaverse, such as Millennials or Generation X. Therefore, it is possible to identify at least three core audiences:

• Rescue or emergency personnel.
• Administrative personnel belonging to agencies that directly supervise the behaviour and problems arising from this gas.
• Educators and recreational supervisors are tasked with conducting awareness and orientation sessions on the subject.

Interest could also correspond to specific communities involved in the problem, although this may eventually require additional digital literacy actions, all subject to the connection infrastructure itself. However, this scenario opens another problem and point for reflection: a metaverse that is more technical and focused on training operational personnel and communities requires the involvement not only of technology companies and specialized organizations but also of public administrations at all levels: local, regional, and national.

Conclusions

As in video games, in the metaverse—especially gamified or immersive proposals—the user can't procrastinate or engage in multiscreen behaviour since the content demands his full attention and involvement. The development of a virtual experience that effectively engages younger users, who are susceptible to an ephemeral communication model and a high rate of digital multimedia stimuli, can allow the message to reach this target audience.

Awareness campaigns are not only necessary but also need to be constantly reviewed and updated, especially if the scope of action is the Internet. The hyper-segmentation of audiences has made any attempt at "mass" communication more complex without the use of prominent prescribers, which implies higher expenses for recruitment purposes.

The metaverse should not be seen as a fad or an instrument of astonishment for certain audiences. The virtual option, and in this specific case, the one that implies interaction and socialization, must be considered as one more alternative to trying to increase the message. Therefore, it is necessary to have very precise knowledge about the metaverse to be used and the audiences that are concentrated there. And see that this fits in with the immediate communication objectives. Moreover, the metaverse should never be seen as a massive option. Depending on the metaverse and its connection options (narrative, devices, etc.), the audiences to be considered will be—according to current considerations—scarce and very characteristic, but as already stated, it should not be dismissed for that reason.

Radon gas, like many other problems affecting public health, could find in the metaverse an ideal channel to connect in a more direct and closer way with the desired audiences. The possibility of contravening any law of physics in the virtual realm endows this alternative with an enormous amount of narrative, aesthetic, and discursive possibilities. However, it will be the obligation of public administrations to promote not only this type of disruptive communication strategy but also to undertake effective digital literacy actions that allow a greater number of social sectors not only to have access to these technological resources but also to find meaning in their use for the purposes that have been presented here.

Nevertheless, information campaigns utilizing new and impactful technologies, such as the metaverse, should target younger audiences in various areas like education and entertainment. This approach aims to empower children and adolescents to serve as a transmission belt of knowledge to older individuals on crucial matters, including prevention against exposure to ionizing radiation that can lead to radon gas.

From the private sector, there are already initiatives to use the metaverse as a mechanism for expansion and increased outreach, as in the case of the ventilation systems company Siber (see *https://shorturl.at/puwAS*), which not only expands its marketing possibilities but also applies new demonstration methods, including explanations on the treatment of radon gas. Inevitably, synergies and joint efforts with companies and individuals must be sought from the public sector.

References

Abbott, C. (2007). Cyberpunk cities: Science fiction meets urban theory. *Journal of Planning Education and Research, 27*(2), 122–131.

Accenture (2022). Technology vision 2022. *Nos vemos en Metaverso.* www.accenture.com/content/dam/accenture/final/a-com-migration/manual/r3/pdf/Accenture-Technology-Vision-2022-Nos-Vemos-En-El-Metaverso.pdf

Barros-Dios, J. M., Ruano-Ravina, A., Perez-Rios, M., Castro-Bernárdez, M., Abal-Arca, J., & Tojo-Castro, M. (2012). Residential radon exposure, histologic types, and lung cancer risk. A case–control study in Galicia, Spain. *Cancer Epidemiology, Biomarkers & Prevention, 21*(6), 951–958. https://doi.org/10.1158/1055-9965.EPI-12-0146-T

Baum, P., Winter, H., Eichhorn, M., Roesch, R., Taber, S., Christopoulos, P., Wiegering, A., & Lenzi, J. (2022). Trends in age—and sex-specific lung cancer mortality in Europe and Northern America: Analysis of vital registration data from the WHO Mortality Database between 2000 and 2017. *European Journal of Cancer, 171*, 269–279.

Baysal, S. (2014). Working with generations X and Y in generation Z period: Management of different generations in business life. *Mediterranean Journal of Social Sciences, 5*(19), 218–229.

BOE. (2022, December 20). Real Decreto 1029/2022, de 20 de diciembre, por el que se aprueba el Reglamento sobre protección de la salud contra los riesgos derivados de la exposición a las radiaciones ionizantes. *Ministerio de la Presidencia, Relaciones con las Cortes y Memoria Democrática.* Referencia: BOE-A-2022–21682.

Bonelo, M. K., & Amar, R. V. (2023). Fortnite, videojuego y educación primaria. *DIM: Didáctica, Innovación y Multimedia, 2023, 41.*

Castronova, E. (2001). Virtual Worlds; a first-hand account of the market and society of the Cyberian frontier. *CESinfo Working Papers, 618*,1–40.

Changhee, K. (2021). Smart city-based metaverse study on the solution of urban problems. *Journal Chosun Natural Science, 14*(1), 21–26.

Dede, C. (1995). The evolution of constructivist learning environments. *Educational Technology, 35*(5), 46–52.

Do Kyun, K., & Kreps, G. L. (2020). An analysis of government communication in the United States during the COVID-19 pandemic: Recommendations for effective government health risk communication. *World Med Health Policy, 12*(4), 398–412.

Drake, B. E., Strelzoff, A., & Sulbaran, T. (2011). Teaching marketing through a micro-economy in virtual reality. *Journal of Marketing Education, 33*(3), 295–311. www.doi.org/10.1177/0273475311420236

Fernández, I. (2022, October 14). El control de España sobre el radón llega "tarde" y sin plan nacional. *Redacción Médica.* www.redaccionmedica.com/secciones/neumologia/el-control-de-espana-sobre-el-radon-llega-tarde-y-sin-plan-nacional--3443

Freudenstein, F., Croft, R. J., Wiedemann, P. M., Verrender, A., Böhmert, C., & Loughran, S. P. (2020). Framing effects in risk communication messages—Hazard identification vs. risk assessment. *Environmental Research, 190*. https://doi.org/10.1016/j.envres.2020.109934.

García-Orosa, B., & Pérez-Seijo, S. (2020). The use of 360° video by international humanitarian aid organizations to spread social messages and increase engagement. *VOLUNTAS: International Journal of Voluntary and Nonprofit Organizations, 31*, 1311–1329. https://doi.org/10.1007/s11266-020-00280-z

Godoy, M. (2023, April 2). Renfe quiere ser el primer operador ferroviario europeo en explotar el metaverso: Esta es su hoja ruta para lograrlo. *Business Insider*. www.businessinsider.es/renfe-apuesta-metaverso-hoja-ruta-1224444

Harris, D. J., Arthur, T., Kearse, J., Olonilua, M., Hassan, E. K., De Burgh, T. C., Wilson, M. R., & Vine, S. J. (2023). Exploring the role of virtual reality in military decision training. *Frontiers in Virtual Reality, 4*, 1165030. www.doi.org/10.3389/frvir.2023.1165030

Herranz de La Casa, J. M., & Bautista, P. S. (2023). From the 360° photo to the metaverse: Conceptual and technical evolution of virtual and immersive journalism from Spain. *Brazilian Journalism Research, 19*(2), 562. https://doi.org/10.25200/BJR.v19n2.2023.1562

Heydari, S. T., Zarei, L., Sadati, A. K., Moradi, N., Akbari, M., Mehralian, G., & Bagheri, K. (2021). The effect of risk communication on preventive and protective behaviours during the COVID-19 outbreak: Mediating role of risk perception. *BMC Public Health, 21*, 54. https://doi.org/10.1186/s12889-020-10125-5

Horowitz, B. (2022, December 15). How AR & VR in healthcare enhances medical training. *HeatlhTech*. https://healthtechmagazine.net/article/2022/12/ar-vr-medical-training-2023-perfcon

INE. (2022). *La Encuesta sobre Equipamiento y Uso de Tecnologías de Información y Comunicación en los Hogares*. Instituto nacional de Estadística, España. www.ine.es/dyngs/INEbase/es/operacion.htm?c=Estadistica_C&cid=1254736176741&menu=resultados&idp=1254735976608

INE. (2023). *Padrón de población a 1 de enero de 2022*. www.ine.es/jaxi/Tabla.htm?path=/t20/e245/p08/l0/&file=03002.px&L=0

Jakma, A. (2022, December 7). *How brands are leveraging the metaverse for awareness*. www.linkedin.com/pulse/how-brands-leveraging-metaverse-awareness-alex-jakma/

JP Morgan. (2022). *Opportunities in the metaverse how businesses can explore the metaverse and navigate the hype vs. reality*. www.jpmorgan.com/content/dam/jpm/treasury-services/documents/opportu-nities-in-the-metaverse.pdf

Lane, R. S., Tomášek, L., Zablotska, L. B., Rage, E., Momoli, F., & Little, J. (2019). Low radon exposures and lung cancer risk: Joint analysis of the Czech, French, and Beaverlodge cohorts of uranium miners. *The International Archives of Occupational and Environmental Health, 92*, 747–762.

Loke, S. -K. (2015). How do virtual world experiences bring about learning? A critical review of theories. *Australasian Journal of Educational Technology, 31*(1). https://doi.org/10.14742/ajet.2532

Lu, J. (2020). Themes and evolution of misinformation during the early phases of the COVID-19 outbreak in China—An application of the crisis and emergency risk communication model. *Frontier in Communication, 5*.

Lu, M. (2021, June 22). How media consumption evolved throughout COVID-19. *Visual Capitalist*. https://bit.ly/3qsRs6T

Malecki, K., Keating, J. A., & Safdar, N. (2021). Crisis communication and public perception of COVID-19 risk in the era of social media. *Clinical Infectious Diseases, 72*, 697–702.

McCrindle, M., & Fell, A. (2021). *Generation alpha*. Hachette.

Ministerio de Sanidad. (2021). *Acción frente al radón* (Colección estudios, informes e investigación) Ministerio de Sanidad.

Moioli, F. (2022, August 11). The Metaverse: Don´t confuse it with virtual reality. *Forbes*. www.forbes.com/sites/forbestechcouncil/2022/08/11/the-metaverse-dont-confuse-it-with-virtual-reality/

Nafarrete, J. (2017, August 23). KFC has a VR job training simulator for new employees. *VRScout*. https://vrscout.com/news/kfc-vr-job-training-simulator-new-employees/

Negreira-Rey, M. C., & Vázquez-Herrero, J. (2022). La cobertura mediática sobre el gas radón en los medios digitales en Galicia. *Prisma Social: Revista de Investigación Social, 39*, 4–24.

Ngoc, L. T. N., Park, D., & Lee, Y. -C. (2023). Human health impacts of residential radon exposure: Updated systematic review and meta-analysis of case—Control studies. *International Journal Environmental Research and Public Health, 20*. https://doi.org/10.3390/ijerph20010097

Poland, G. A., & Spier, R. (2010). Fear, misinformation, and innumerates: How the Wakefield paper, the press, and advocacy groups damaged the public health. *Vaccine, 28*(12), 2361–2362.

Prensky, M. (2001). Digital natives, digital immigrants. *On The Horizon, 9*(5), 1–6.

Ruano-Ravina, A., Dacosta-Urbieta, A., Barros-Dios, J. M., & Kelsey, K. (2018). Radon exposure and tumors of the central nervous system. *Gaceta Sanitaria, 6*, 567–575.

Serino, S., & Repetto, C. (2018). New trends in episodic memory assessment: Immersive 360° ecological videos. *Frontiers in Psychology, 9*, 1–6. https://doi.org/10.3389/fpsyg.2018.01878

Sidorenko B. P., & Herranz, J. M. (2021, October 28). Un viaje por los universos virtuales (y publicitarios) del metaverso. *The Conversation*. https://bit.ly/3t98lqj

Soto de la Cruz, J., de la Hera, T., Cortés Gómez, S., & Lacasa, P. (2023). Digital games as persuasion spaces for political marketing: Joe Biden's campaign in fortnite. *Media and Communication, 11*(2), 266–277.

Torres-Durán, M., Ruano-Ravina, A., Parente-Lamelas, I., Leiro-Fernández, V., Abal-Arca, J., Montero-Martínez, C., et al. (2014). Lung cancer in never-smokers: A case-control study in a radon-prone area (Galicia, Spain). *The European Respiratory Journal, 44*, 994–1001.

Turner, A. (2015). Generation Z: Technology and social interest. *The Journal of Individual Psychology, 71*(2), 103–113.

Ulmer, J., Braun, S., Cheng, C., Dowey, S., & Wollert, J. (2020). Gamified virtual reality training environment for manufacturing industry. *19th International Conference on Mechatronics*. www.doi.org/10.1109/ME49197.2020.9286661

Ventura, S., Brivio, E., Riva, G., & Baños, R. (2019). Immersive versus non-immersive experience: Exploring the feasibility of memory assessment through 360° technology. *Frontiers in Psychology, 10*(2509), 1–7. https://doi.org/10.3389/fpsyg.2019.02509

Vergara, V., Caudell, T., Goldsmith, P., & Alverson, D. (2008). Knowledge driven design of virtual patient simulations. *Innovate, 5*(2), 1–6.

Victoria González, V. (2020). Herramientas TIC para la gamificación en Educación Física. *Edutec. Revista Electrónica De Tecnología Educativa, 71*, 67–83. https://doi.org/10.21556/edutec.2020.71.1453

WHO. (2009). *WHO handbook on indoor radon: A public health perspective*. Geneva.

WHO. (2021, February 2). *El radón y sus efectos en la salud*. www.who.int/es/news-room/fact-sheets/detail/radon-and-health

YPulse. (2022, March 15). *Here's what Gen Z & Millennials are buying in the metaverse*. https://bit.ly/36Arrw

Part III
Public perception of risk

7 When risk is invisible and at home

News coverage on radon gas in local media

María-Cruz Negreira-Rey,
Jorge Vázquez-Herrero and Rita Araújo

Introduction

Radon is a colourless, odourless and tasteless radioactive gas. It is inherently present in nature, as it originates from the disintegration of uranium, an element found in the ground and rocks. As the title of this chapter indicates, it 'is invisible and at home', so radon is considered a relevant risk factor, since it is an important source of radiation for the general population. As a result of this radiation, radon causes health problems; in fact, since 1988, it has been recognized as a human carcinogen (International Agency for Research on Cancer, 1988), specifically related to lung cancer, being the second most frequent cause for tobacco users and the first cause of the disease in non-smokers.

Its ubiquitous condition, although with different levels of concentration around the world, makes the spaces we inhabit the most worrying due to the time of exposure to the gas, for example, in the home and the workplace. This situation has required regulation and measures to be taken, initially for the protection of workers in specific sectors and subsequently for buildings and habitability. Measurement in enclosed spaces is performed with passive detectors and observing the average, according to the protocols established in each country, as concentrations vary circumstantially. The reference value, which is considered the upper limit of concentration for dwellings in different regulatory frameworks, is 100 Bq/m3 in general or 300 Bq/m3 in countries with particularities (World Health Organization, 2009).

In this chapter, the news coverage of radon gas in the local media in Spain will be addressed. Here, the national reference data are provided by the Nuclear Safety Council in the predictive map of radon exposure (García-Talavera et al., 2013), the map of natural gamma radiation (Suárez et al., 2000) and the mapping of radon potential (García-Talavera & López, 2019). Mapping is an indispensable resource for designing measures in the most exposed areas. The mapping of radon potential in Spain has established priority action areas: those in which 10% of the buildings have concentrations above the level of 300 Bq/m3 (García-Talavera & López, 2019). These areas are identified as representing

DOI: 10.4324/9781032618180-10

17% of the national territory, with the communities of Galicia, Extremadura and Madrid seeing more than 30% of their surface area affected (Nuclear Safety Council, 2017).

Regarding research on communication and radon gas, some previous studies have focused on the role of news media. Brewster (2015) analysed Canadian newspaper coverage of radon, concluding that it responded to developments in governmental regulation and research. He further noted that there was no clear authority and that there was no correlation between the volume of coverage and exposure to the gas, public awareness or health significance. Post (1986) focused on the role of local newspapers and warned of the difficulty of covering radon: it is complex, there are economic pressures and it requires specialist journalists. Among his conclusions, he noted that non-scientific governmental sources were favoured. Negreira-Rey and Vázquez-Herrero (2022) analysed news coverage in digital media in Galicia—the most affected region in Spain—and identified that the main focuses are 'health and prevention', 'research' and 'housing and urban planning'. The geographical scope of the news is mostly local, and the sources are mainly research institutions and public administration.

Nowadays, radon risk communication strategies go beyond newspapers to websites, social media, conferences, journals, magazines, seminars, maps, workshops and beyond (Cheng, 2016; Ferguson & Valenti, 1988; Makedonska et al., 2018). This communication is largely institutional in nature, so researchers pay attention to governmental programmes. Lofstedt (2019) studied radon risk communication at the Swedish National Board of Housing and Planning, concluding that specific budgets are necessary to run effective campaigns and achieve government goals. Page (1994) analysed the EPA's Radon Program (United States) and called for communication that persuades people to understand there is a personal responsibility for radon prevention and encourages a dialogue between the scientific community and the public. In Bulgaria, Makedonska et al. (2018) focused on the National Radon Program, while the Regional Prevention Plan in Sardinia (Italy) was analysed by Cori and colleagues (2022), who called for more science-based communication programmes. This concern for efficient communication is reflected in the 2021 publication of a manifesto to guide radon communication strategy (Bouder et al., 2021), which includes several recommendations, such as re-framing radon from 'a natural radioactive gas' to 'indoor air pollution', the use of interactive tools, support for social science research on radon and the maintenance of continuous communication, among other considerations.

Considering this background, both the relevance of radon and its risk communication and scientific interest, this chapter aims to understand the relevance of radon gas in the media agenda of local digital media in Spain, analysing its evolution in recent decades and the characteristics of the news coverage. Complementing this objective is the identification of the most relevant actors in the information, observing the role of public administrations and political actors in the communication of the risk associated with radon. The following sections provide a brief theoretical background on health and risk communication, a methodological overview of the research that supports the chapter, a section on results and trends observed in the coverage of radon gas in the Spanish local media and some concluding thoughts.

Health and risk communication

Health communication has emerged as one of the most important public health issues in this century (Ishikawa & Kiuchi, 2010)—its key goals are "to engage, empower, and influence individuals and communities" (Schiavo, 2014). This refers to the array of communication processes and messages that are constituted around health issues (Dutta & Zoller, 2008). There is, nonetheless, a broad awareness of the scope of health communication and its strategic areas since health is global and cuts across society. The recent outbreak of SARS-CoV-2 highlighted the importance of health communication, risk communication and community engagement during a public health crisis, which usually concerns the outbreak of particular diseases or the identification of specific risks from environmental or lifestyle factors (Reynolds & Seeger, 2005). Since emerging diseases or hazards create high levels of uncertainty, communication becomes central in managing a public health emergency. Therefore, "public health communication is the scientific development, strategic dissemination, and critical evaluation of relevant, accurate, accessible, and understandable health information communicated to and from intended audiences to advance the health of the public" (Bernhardt, 2004, p. 2051). "Preventing, controlling, and managing risk is one of the central goals of health communication. As a result, risk communication has developed as an area of research and practice closely related to health communication" (Lopez & Cho, 2023), hence risk communication is a well-established field at the centre of many public health campaigns.

Despite risk being a "very sophisticated math concept that has proven problematic to define" (Rudd & Baur, 2020), risk communication presents health messages that seek to promote behavioural change. It is "grounded in an assumption that the public has a generalized right to know about hazards and risks" since providing people with information allows them to make informed choices regarding risk (Reynolds & Seeger, 2005, p. 45). In the aftermath of the COVID-19 pandemic, Ratzan, Sommarivac and Rauh considered that the current characteristics of the media environment and the way people engage with the news "call for a revision of the risk communication guidance during a public health crisis" (2020, p. 2), given that "understanding and managing risk is central for bolstering individual and public health" (Lopez & Cho, 2023). Moreover, risk management may use several tools, such as risk maps, local capacity-building or universal measures that limit social contact, but they all need to be effectively defined, coherent and implemented and communicated to populations (Teixeira et al., 2021). Due to its complexity, effective communication about public health issues is a challenge, even without a health crisis (Vraga & Jacobsen, 2020).

The news media are often the primary source of health information for the lay public, and health and medicine have long been prominent topics in the media (Diviani et al., 2023; Hallin & Briggs, 2014). Indeed, communication is a primary public health strategy; it helps people stay up-to-date and make informed decisions regarding their own health (Garfin et al., 2020; Rudd & Baur, 2020), as well as being prepared to take action, "especially during times of natural disasters and emergencies" (Rudd & Baur, 2020). Indeed, during a health crisis, the public depends on the media to convey accurate and up-to-date information (Garfin et al., 2020). Not surprisingly, the COVID-19 pandemic has reinforced the value of accurate and reliable information, resulting in greater confidence in the media (Newman et al., 2021) and growth in news consumption (Casero-Ripollés, 2020; MacDonald, 2021). Despite recent data showing that trust in news has fallen, "reversing in many countries the gains made at the height of the Coronavirus pandemic" (Newman et al., 2023), the news media continue to be particularly important in processing (health) news, playing a central role in "disseminating information about a crisis, highlighting key incidences and holding decision-makers to account for their actions" (McGuire et al., 2020, p. 364). Health news is a very broad category that includes stories on diseases, new treatments or services, health products or procedures, medical research and much more (Araújo, 2023).

Since journalists covering the health beat often have little or no formal education in the health field, they rely heavily on news sources, including official and expert ones (Araújo, 2023). This is particularly important given that health news may contribute to one's decision-making process. Thus—especially during a public health crisis—communication should be clear, honest and consistent (Finset et al., 2020; Garfin et al., 2020; Ratzan et al., 2020). Furthermore, journalists should ensure that media coverage is evidence-based and accurate (Schiavo, 2020); they can remind the public about important preventive measures, such as staying home, washing hands or wearing masks (Velázquez & Serna-Zamarrón, 2020). Information should be updated when relevant by a designated spokesperson who should acknowledge that recommendations may change whenever new evidence emerges (Finset et al., 2020).

The media play a central role in the public response to a pandemic, working as mediators between governments, health institutions and the public (Mheidly & Fares, 2020). Working with the media to help them understand science will improve risk communication to the public and reduce inappropriate concerns and panic (McCloskey & Heymann, 2020). This was recognised during the COVID-19 pandemic, and the response to this global outbreak depended largely on public trust in government officials (Nutbeam, 2020). In health and risk communication, trust is essential—"once lost, trust and credibility are nearly impossible to regain" (Fielding, 2020). And even though "expert sources may be considered highly credible in one sense (expertise)", they are often not "especially credible in another (trustworthiness)" (Geiger, 2020). New Zealand, for instance, was pointed out as a successful example of managing risk communication during the pandemic because official communication was centralized in one government official and the Director-General of Health, who provided clear messages with simple instructions to promote preventive behaviours (Jamieson, 2020). "Having a clear voice within government helps avoid a 'talking heads' dynamic that undermines the development of a cohesive strategy" (Ratzan et al., 2020, p. 3) and the reliance on less trustworthy sources may open the door to mis- and dis-information (Dunwoody, 2020).

An approach to news coverage on radon gas

This study of radon gas coverage in the local news media in Spain took as a sample the digital editions of 23 local media outlets from the provinces and autonomous communities most affected by the gas Andalucía, Asturias, Castilla-La Mancha, Castilla y León, Comunidad de Madrid, Galicia and Extremadura. The research is based on a content analysis of news published in the period 2002–2022, identified by searching for the word 'radón' in the archive of the digital media. A total of 1,049 news items were retrieved, from which a smaller sample was selected for the content analysis, for a total of 579 items, with a confidence level of 95% and a margin of error of 5%. News content was analysed by applying a sheet in which the basic identifying data of each piece, the main topic (radon or others), the thematic focus, the sources of information and the scope of the news item were collected. The description and magnitude of the risk associated with radon gas in the news items were also recorded.

The analysis reveals that news coverage on radon gas in Spain has grown progressively over the last two decades. The number of news items on radon notably increased from 2017 onwards, with an average of 64.7 news items published each year—2018 shows a peak in the number of news items, with 103 published. In the last few years—with a greater volume of published information on radon—coincided with some advances in the regulatory framework. In 2018, Directive 2013/59/Euratom should have been enacted, by which the European Union urged Member States to establish exposure limits and monitor radon gas levels in workplaces and enclosed spaces, to define strategies for action against radon and for communication to the public. In Spain, Royal Decree 732/2019 was approved to amend the Technical Building Code and update prevention measures in new and existing buildings. At the end of 2022, Royal Decree 1029/2022 on health protection against risks derived from exposure to ionising radiation was approved, repealing the previous text of 2001. Other previous regulatory measures include Royal Decree 314/2006, which updated the regulations on the presence of radon and the quality of water for human consumption (Ministerio de Sanidad et al., 2021).

In addition to knowing the volume of news, it is necessary to observe the approach and treatment of the information on radon that reaches the public via news media. Of the sample analysed, 49.7% of the news items have radon as their main topic. Major thematic approaches cover health and prevention, research, housing and urban planning and the regulatory framework and policies. Radon is a newsworthy topic mainly because of the risk it poses to people's health, as its exposure is directly linked to the likelihood of developing cancer, particularly lung cancer. This is the rationale underpinning prevention and regulatory measures against radon gas.

One of the main topics in the news coverage of radon is its measurement in order to discover the most affected areas and territories. News reports inform the public about measurement plans at national, regional and local levels, in which organisations such as the Spanish Nuclear Safety Council, regional and municipal governments or specialised research groups are involved. In the field of local journalism, news reporting on monitoring measures and their results at the municipal, regional, provincial or autonomous community levels is more frequent. In addition to measuring the territorial incidence of radon—in soil and in some cases, also in water—at the local level, reports on measurements in workplaces, schools or homes are frequent.

Progress and new knowledge on the presence, measurement, effects and solutions against radon gas reach news media, for the most part, through the communication and dissemination of results by research groups that specialise in the subject. In Spain, there are outstanding research teams, such as the Radon Group of the LARUC Laboratory of the University of Cantabria or the Galician Radon Laboratory of the University of Santiago de Compostela, which regularly update measurement maps and disseminate these and other results through the media. The research and voices of the experts are also present in the news coverage of conferences or dissemination talks for professionals and the general public.

Knowledge of radon levels and its effects demands action. In local media reporting on radon, news about renovation of public buildings, relocation of workers to free them from spaces with high radon levels, information on technical solutions for housing or the general recommendation to air out buildings in at-risk areas can be found. Specific solutions for public buildings often come from government bodies, as well as incentives to encourage the measurement and implementation of measures in private properties—for example, the Xunta de Galicia with an economic support plan for solutions in housing or the Junta de Extremadura with a guide of good practices for building. Specialised companies also have a voice in the news, albeit less frequently, providing information on their solutions for radon gas measurement and removal.

Regulation of radon is also one of the issues highlighted in the news coverage. One of the rules that attracted the most media attention was the aforementioned Directive 2013/59/Euratom to explain its progress but especially to highlight the lack of action and compliance on the part of the Spanish government. The other measure most reported in the media is Royal Decree 732/2019, which updates the Technical Building Code to include measures to be taken against radon in new constructions. At the local level, some measures taken by regional governments are highlighted, such as financial aid plans for the implementation of solutions and the rehabilitation of buildings or good practice guides.

Although the thematic approaches of news items on radon may vary, a more complete treatment of the information with a contextual and explanatory character is observed, especially in the longer pieces. Current data on radon in the news are often complemented with an explanation of what radon is, why it originates, its relationship with the risk of cancer, the level of exposure of the territory and possible solutions or the regulatory framework.

Among the news items with radon as a secondary topic, the most frequent are those dealing with cancer, its typologies and risk factors. Also, those concerning urban planning, occupational safety, municipal politics—when radon appears as a matter of debate or decision—or the environment—mainly in relation to geological studies or actions.

The most frequently cited sources of information are those belonging to the public administration, research bodies and institutions, experts and political parties. These sources dominate the news agenda and set the topics and approaches of the information, as mentioned previously. In the news, it is very rare to find the voice of civil associations or citizens giving their personal testimony.

Regarding the degree of proximity of the information, it is observed that the municipal and regional scope dominates the news coverage. The news on radon gas reports on what is happening in the areas closest to the citizens, a key issue in order to know the possible risks to which they are exposed and the measures and solutions proposed by the public administration.

Conclusion

News media play a fundamental role in communicating risk to the public. Their reach and capacity to influence make them a favourable channel to inform, explain, raise awareness or warn about a given risk, in this case, about the health hazards of exposure to radon gas (Perko, 2012). However, the media construct their news coverage based on criteria of newsworthiness, the relevance of certain sources of information and based on particular approaches and frameworks (Lichtenberg & MacLean, 1991).

The analysis of the news coverage of radon gas in the local media in Spain reveals some key findings. The intensity of radon coverage is increasing, with a steady growth in the number of news items on the subject, while research, regulation and knowledge of its effects and solutions are also advancing. In addition to news stories, explanatory and contextual information is often included in the news. It is necessary for the public to know what radon is and how it originates, the incidence of the gas in their territory, the possible risks it may have on people's health, how it can be removed from their homes or other buildings or what measures are proposed by government bodies. Communicating about a risk that has a permanent natural origin that is not perceived by the senses and has medium- or long-term effects is complex and requires this didactic and awareness-raising function of journalism.

The research also reveals that government and public administration sources, institutional bodies and experts dominate the news coverage of radon. They are the actors that produce relevant advances in knowledge and regulation about the gas and are, therefore, the most cited in the news. The dissemination of new findings or the announcement of measures to implement solutions against the presence of gas are important to transmit capacity for action, safety and the possibility of reversing the risk to which citizens may be exposed.

The local character of the news can play a key role in the perception of risk. Just like novelty, relevance of sources, risk or conflict, proximity is also a newsworthy value in radon news. The news becomes more important for citizens when it informs them about regulatory changes that affect them directly, such as the incidence of gas in their territory or in areas they frequent. Feeling the risk close to them can determine their perception of the danger to which they are exposed and the effects on their health that they could suffer. This is why the municipal and regional focus is predominant in the news items analysed, although the national and international context is also present, especially regarding the regulatory framework.

In further research, it will be important to delve deeper into citizens' perception of news about radon gas to analyse its effects on the awareness of risk. The news reports on radon gas from local sources can be decisive in motivating citizens to take action and find solutions, but it is necessary to address this issue through further academic research.

Acknowledgments

This chapter is part of the project *Radon in Spain: Public perception, media agenda and risk communication* (RAPAC), financed by the Spanish Nuclear Safety Council [Consejo de Seguridad Nuclear] (SUBV-13/2021).

References

Araújo, R. (2023). Media content, news. In E. Y. Ho, C. L. Bylund, J. C. M. van Weert, I. Basnyat, N. Bol, & M. Dean (Eds.), *The international encyclopedia of health communication* (pp. 1–5). Wiley Blackwell. https://doi.org/10.1002/9781119678816.iehc0629

Bernhardt, J. M. (2004). Communication at the core of affective public health. *American Journal of Public Health, 94*(12), 2051–2053. https://doi.org/10.2105/ajph.94.12.2051

Bouder, F., Perko, T., Lofstedt, R., Renn, O., Rossmann, C., Hevey, D., Siegrist, M., Ringer, W., Pölzl-Viol, C., Dowdall, A., Fojtíková, I., Barazza, F., Hoffmann, B., Lutz, A., Hurst, S., & Reifenhäuser, C. (2021). The Potsdam radon communication manifesto. *Journal of Risk Research, 24*(7), 909–912. https://doi.org/10.1080/136698 77.2019.1691858

Brewster, L. (2015). *Radon gas portrayal in the Canadian print media: A mixed methods approach* (Degree of Master, Simon Fraser University). http://summit.sfu.ca/ item/15261

Casero-Ripollés, A. (2020). Impact of COVID-19 on the media system. Communicative and democratic consequences of news consumption during the outbreak. *Profesional de la Información, 29*(2), e290223. https://doi.org/10.3145/epi.2020.mar.23

Cheng, W. (2016). Radon risk communication strategies: A regional story. *Journal of Environmental Health, 78*(6), 102–107. www.jstor.org/stable/26330399

Cori, L., Bustaffa, E., Cappai, M., Curzio, O., Dettori, I., Loi, N., Nurchis, P., Sanna, A., Serra, G., Sirigu, E. Tidore, M., & Bianchi, F. (2022). The role of risk communication in radon mapping, risk assessment and mitigation activities in Sardinia (Italy). *Advances in Geosciences, 57*, 49–61. https://adgeo.copernicus.org/articles/57/49/2022/

Diviani, N., Rubinelli, S., & Fiordelli, M. (2023). Media, quality of health information. In E. Y. Ho, C. L. Bylund, J. C. M. van Weert, I. Basnyat, N. Bol, & M. Dean (Eds.), *The international encyclopedia of health communication*. Wiley Blackwell. https://doi. org/10.1002/9781119678816.iehc0630

Dunwoody, S. (2020). Science journalism and pandemic uncertainty. *Media and Communication, 8*(2), 471–474. https://doi.org/10.17645/mac.v8i2.3224

Dutta, M. J., & Zoller, H. M. (2008). Theoretical foundations. Interpretative, critical, and cultural approaches to health communication. In H. M. Zoller & M. J. Dutta (Eds.), *Emerging perspectives in health communication* (pp. 1–28). Routledge.

Ferguson, M. A., & Valenti, J. M. (1988). *Risk-taking tendencies and radon messages: A field experiment testing an information processing model for risk communication.* https://eric.ed.gov/?id=ED298545

Fielding, J. (2020, March 3). Good communication will help beat COVID-19. *The Hill*, October 11, 2023. https://thehill.com/opinion/healthcare/490410-good-comm unications-will-help-beat-covid-19

Finset, A., Bosworth, H., Butow, P., Gulbrandsen, P., Hulsman, R. L., Pieterse, A. H., Street, R., Tschoetschel, R., & Van Weert, J. (2020). Effective health communication— A key factor in fighting the COVID-19 pandemic. *Patient Education and Counseling, 103*(5), 873–876. https://doi.org/10.1016/j.pec.2020.03.027

García-Talavera, M., & López, F. J. (2019). Cartografía del potencial de radón de España. *Consejo de Seguridad Nuclear*, September 1, 2023. www.csn.es/docu ments/10182/27786/INT-04.41+Cartograf%C3%ADa+del+potencial+de+rad%C3%B 3n+de+Espa%C3%B1a

García-Talavera, M., Martín, J. L., Gil, R., García, J. P., & Suárez, E. (2013). El mapa predictivo de exposición al radón en España. *Consejo de Seguridad Nuclear*, September 1, 2023. www.csn.es/documents/10182/27786/INT-04-31%20El%20 mapa%20predictivo%20de%20exposici%C3%B3n%20al%20rad%C3%B3n%20 en%20Espa%C3%B1a

Garfin, O. R., Silver, R. C., & Holman, E. A. (2020). The novel coronavirus (COVID-19) outbreak: Amplification of public health consequences by media exposure. *Health Psychology, 39*(5), 355–357. https://doi.org/10.1037/hea0000875

Geiger, N. (2020). Do people actually "listen to the experts"? A cautionary note on assuming expert credibility and persuasiveness on public health policy advocacy. *Health Communication, 37*(6), 677–684. https://doi.org/10.1080/10410236.2020.1862449

Hallin, D. C., & Briggs, C. L. (2014). Transcending the medical/media opposition in research on news coverage of health and medicine. *Media, Culture & Society, 37*(1), 85–100. https://doi.org/10.1177/0163443714549090

International Agency for Research on Cancer. (1988). *IARC monographs on the evaluation of the carcinogenic risks to humans* (p. 43). World Health Organization. https://publications.iarc.fr/Book-And-Report-Series/Iarc-Monographs-On-The-Identification-Of-Carcinogenic-Hazards-To-Humans/Man-Made-Mineral-Fibres-And-Radon-1988

Ishikawa, H., & Kiuchi, T. (2010). Health literacy and health communication. *Biopsycho Social Medicine, 4*(1), 18. https://doi.org/10.1186/1751-0759-4-18

Jamieson, T. (2020). "Go hard, ho early": Preliminary lessons from New Zealand's response to COVID-19. *The American Review of Public Administration, 50*(6–7), 598–605. https://doi.org/10.1177/0275074020941721

Lichtenberg, J., & MacLean, D. (1991). The role of the media in risk communication. In R. E. Kasperson & P. J. M. Stallen (Eds.), *Communicating risks to the public. Technology, risk, and society*. Springer. https://doi.org/10.1007/978-94-009-1952-5_9

Lofstedt, R. (2019). The communication of radon risk in Sweden: Where are we and where are we going?. *Journal of Risk Research, 22*(6), 773–781. https://doi.org/10.1080/13669877.2018.1473467

Lopez, R., & Cho, H. (2023). Risk communication. In E. Y. Ho, C. L. Bylund, J. C. M. van Weert, I. Basnyat, N. Bol, & M. Dean (Eds.), *The international encyclopedia of health communication*. Wiley Blackwell. https://doi.org/10.1002/9781119678816.iehc0979

MacDonald, N. E. (2021). COVID-19, public health and constructive journalism in Canada. *Canadian Journal of Public Health, 112*(2), 179–182. https://doi.org/10.17269/s41997-021-00494-8

Makedonska, G., Djounova, J., & Ivanova, K. (2018). Radon risk communication in Bulgaria. *Radiation Protection Dosimetry, 181*(1), 26–29. https://doi.org/10.1093/rpd/ncy096

McCloskey, B., & Heymann, D. L. (2020). SARS to novel coronavirus—old lessons and new lessons. *Epidemiology and Infection, 148*, e22. https://doi.org/10.1017/S0950268820000254

McGuire, D., Cunningham, J. E. A., Reynolds, K., & Matthews-Smith, G. (2020). Beating the virus: An examination of the crisis communication approach taken by New Zealand Prime Minister Jacinda Arden during the COVID-19 pandemic. *Human Resource Development International, 23*(4), 361–379. https://doi.org/10.1080/13678868.2020.1779543

Mheidly, N., & Fares, J. (2020). Leveraging media and health communication strategies to overcome the COVID-19 infodemic. *Journal of Public Health Policy, 41*(4), 410–420. https://doi.org/10.1057/s41271-020-00247-w

Ministerio de Sanidad, Ministerio de Transportes, Movilidad y Agenda Urbana, Ministerio para la Transición Ecológica y el Reto Demográfico, & Consejo de Seguridad Nuclear. (2021). *Acción frente al radón*. Ministerio de Sanidad. www.sanidad.gob.es/ciudadanos/saludAmbLaboral/medioAmbiente/docs/Accion_Radon.pdf

Negreira-Rey, M. -C., & Vázquez-Herrero, J. (2022). La cobertura mediática sobre el gas radón en los medios digitales en Galicia. *Prisma Social, 39*, 4–24. https://revistaprismasocial.es/article/view/4855

Newman, N., Fletcher, R., Eddy, K., Robertson, C. T., & Nielsen, R. K. (2023). *Digital news report 2023*. Reuters Institute for the Study of Journalism, University of Oxford. https://reutersinstitute.politics.ox.ac.uk/digital-news-report/2023

Newman, N., Fletcher, R., Schulz, A., Andi, S., Robertson, C. T., & Nielsen, R. K. (2021). *Digital news report 2021*. Reuters Institute for the Study of Journalism, University of Oxford. https://reutersinstitute.politics.ox.ac.uk/digital-news-report/2021

Nuclear Safety Council. (2017). Cartografía del potencial de radón de España. *Consejo de Seguridad Nuclear*. www.csn.es/documents/10182/914801/FDE-02.17%20 Cartograf%C3%ADa%20del%20potencial%20de%20rad%C3%B3n%20de%20 Espa%C3%B1a

Nutbeam, D. (2020). COVID-19: Lessons in risk communication and public trust. *Public Health Research and Practice, 30*(2), e3022006. https://doi.org/10.17061/phrp3022006

Page, S. D. (1994). Indoor radon: A case study in risk communication. *American Journal of Preventive Medicine, 10*(3), 15–18. https://doi.org/10.1016/S0749-3797(18)30545-2

Perko, T. (2012). The role of mass media and journalism in risk communication. *Journal of Mass Communication and Journalism, 2*(2), 1000e110. https://doi.org/10.417 2/2165-7912.1000e110

Post, J. F. (1986). Reporting on radon: The role of local newspapers. *Environment: Science and Policy for Sustainable Development, 29*(2), 4–45. https://doi.org/10.1080/00 139157.1987.9928854

Ratzan, S. C., Sommarivac, S., & Rauh, L. (2020). Enhancing global health communication during a crisis: Lessons from the COVID-19 pandemic. *Public Health Research & Practice, 30*(2), e3022010. https://doi.org/10.17061/phrp3022010

Reynolds, B., & Seeger, M. W. (2005). Crisis and emergency risk communication as an integrative model. *Journal of Health Communication, 10*(1), 43–55. https://doi. org/10.1080/10810730590904571

Rudd, R. C., & Baur, C. (2020). Health literacy and early insights during a pandemic. *Journal of Communication in Healthcare, 13*(1), 13–16. https://doi.org/10.1080/1753 8068.2020.1760622

Schiavo, R. (2014). *Health communication. From theory to practice* (2nd ed). Jossey-Bass.

Schiavo, R. (2020). Vaccine communication in the age of COVID-19: Getting ready for an information war. *Journal of Communication in Healthcare, 13*(2), 73–75. https:// doi.org/10.1080/17538068.2020.1778959

Suárez, E., Fernández, J. A., Baeza, A., Moro, M. C., García, D., Moreno, J., & Lanaja, J. M. (2000). *Proyecto Marna. Mapa de radiación gamma natural*. Consejo de Seguridad Nuclear. www.csn.es/documents/10182/27786/INT-04-02+Proyecto+Marna.+M apa+de+radiaci%C3%B3n+gamma+natural

Teixeira, P. M., Vital, D., Araújo, R., & Gomes, B. M. (2021). Risk communication and community engagement in the COVID-19 pandemic in Portugal. *Acta Médica Portuguesa, 34*(1), 1–2. https://doi.org/10.20344/amp.15145

Velázquez, L. E., & Serna-Zamarrón, A. (2020). Cobertura informativa de la pandemia por COVID-19 en Nuevo León: Liderazgo y periodismo con misión de servicio. *Revista Española de Comunicación en Salud* (Suplemento 1), 186–209. https://doi. org/10.20318/recs.2020.5453

Vraga, E. K., & Jacobsen, K. H. (2020). Strategies for effective health communication during the coronavirus pandemic and future emerging infectious disease events. *World Medical & Health Policy, 12*(3), 233–241. https://doi.org/10.1002/wmh3.359

World Health Organization. (2009). *Handbook on indoor radon: A public health perspective*. World Health Organization. www.who.int/publications/i/item/9789241547673

8 Public perception and treatment of risk communication in Latin America

Lila Luchessi and Pablo Escandón-Montenegro

Introduction

"Risk" is a polysemic and ambiguous term (De Almeida et al., 2009). Those who state that risk is problematic also pose that it is not possible to address it without considering that there is a risk only if a vulnerable situation arises. In the Southern Cone,

> The notion of vulnerability seems to be the most suitable one to understand the transformative impact caused by the new pattern of social development and to capture the higher risk to which a large part of Latin American residents are exposed in the current historical period.
>
> (Pizarro, 2001, p. 9)

The idea of risk entails danger. In the southern part of the Americas, danger is based on the "imminent contingency of evil" (RAE, 2023). That evil is not usually linked to natural phenomena, climate-related disasters, or wars. In the south of the south, that evil can be the result of failed decisions, inadequate policies, and corporate pressures. In this case, risk is caused by mistakes that affect other living conditions. Thus, this creates uncertainty. But this does not always lead to crises. Living each day with this restlessness, waiting for the unexpected to happen, with risk lying in wait, can, by itself, trigger some crises. However, the indiscriminate use of the notions of risk and crisis as if they were synonyms adds to the confusion.

As stated by Mario Riorda, "a crisis requires certainties in communication. It aims at battling uncertainty". In contrast, communicating risk means "preventing, raising awareness, modifying habits or behaviours" (2020, p. 21) to mitigate the uncertainty caused by danger.

This juxtaposition between lack of foresight and naturalization of uncertainty is where prevention tasks (in the case of risk management) and creation of certainty (in the case of crisis management) become more complex. These feelings, linked to individual and social misfortune, are not always the result of threats

DOI: 10.4324/9781032618180-11

to the health of each individual, although they do have an impact on it. In many cases, risk and vulnerability are the consequence of material shortfalls, scarcity, and lack of social benefits. In other cases, they are the consequence of being unable to plan, even in the short term.

Clearly, radon gas is a threat to everyone. Argentina applies AR Guideline 1, "Dosimetry Factors for External and Internal Exposure, Guideline Levels of Radionuclides in Food and Water, and Recommendations for Radon Gas Exposure Control", second revision of 24 June 2022 (Resolution 308 of 2022), which supplements the Basic Standard on Radiation Safety (Presidencia de la Nación Argentina, 2022). Despite the guideline and scientific works that warn about its potential consequences, the media and social agenda are not particularly interested in these forms of contamination and its consequences on the health of the population.

In Ecuador, Executive Decree No. 229 regulating the Organic Law of Energy Efficiency (Ecuador, Ministery of Energy, 2021) states that the purpose of energy efficiency "does not involve reducing the comfort of citizens, but rather promotes awareness of the proper use of energy in the population". Despite these regulations, concerns, and the existence of academic works that bring the topic to the scientific field, there is no evidence of it being included in the media in a way that shows radiation risk in society's discussions, in the media agenda, or in social media interactions.

With a very clear conscience on sanitary actions against viral threats, the other risks to which people are exposed start with the economic situation and have an effect on health. In Latin America, economic instability and the fragility of institutions create risks in other areas and affect the sanitary system and citizens' bodies. Sometimes, decisions on public policies, the State's participation—or lack thereof—in preventing sanitary risks, and media agendas that are far removed from the interests of the population and their health move the discussion on risk communication and potential crises to other territories. During the COVID-19 pandemic, preventive or specific operations were commanded by the State in most countries of the region. Where States had a bigger presence, the impact on the life and health of people was significantly lower than in places where each person did the best they could.

Given its geographic, demographic, and topographic characteristics, South America has sufficient resources to produce successful economic policies. However, external conditions, internal mismanagement, and corporate disputes cause more sanitary risks than viruses, bacteria, and other forms of contamination. Upon an appropriate assessment of the repercussions of economic actions on the environment, we can understand that the risk that arises from mistakes, involuntary errors, and lack of foresight due to short-term decisions is broad and has an impact on health.

Two of the main actions taken by States during the pandemic were related to communication: (1) incorporating scientists as Government advisors and (2) creating preventive communication campaigns to prevent mass infections. The relevance of publicly communicating science was evidenced by contingency-dependent actions, although, in most cases, there was no adequate communication plan in place to reduce the anxiety caused by uncertainty. Perhaps because "the main purpose of a Communication Plan is to turn uncertainty and reactive communication into an organized and proactive process ensuring the best conditions for population health care. It is a roadmap. A map that guides and poses issues" (Riorda, 2020, p. 20).

As to risks in other areas of the State and public affairs, it is not always possible to have a plan to guide their communication. With the constant and imminent feeling that something will go wrong comes a certain habituation and resilience to risk. This may seem interesting if the flexibility of different players to adapt to uncertain situations is considered, but it also means normalizing and being familiar with the lack of organization in crisis management.

One of the most visible issues is that risks that can give rise to crises are not always linked to abstract or unforeseen events. Given the economic and topographic conditions, some actions that affect the environment in a direct and planned manner—and their consequences on health—are not perceived as dangerous because of the remoteness (in both space and time) of living in large territories favoured by nature.

In the Southern Cone, the extension of lands, the farming tradition, and the environmental advantages naturalize the consequences of excessive extraction, as if the dormant risk did not exist or was a product of science fiction. Furthermore, natural resource exploitation with contaminating methods is not usually newsworthy, but it affects the health of some populations and urban groups that are exposed to that contamination and, indirectly, has an impact on vulnerable groups, which makes the imminence of danger real.

Among other practices, the use of toxic agrochemicals that poison water and modify the DNA of foods for mass consumption widely spread autoimmune and more severe diseases in a fair share of the population. Thus, risk is not related to infectiousness or viral dissemination. In those cases, risk is related to using elements to increase productivity to the detriment of health.

The cost-benefit analysis seems to incorporate quality of life and existence in itself. As long as there is a profit, risks brought about by the consumption of or closeness to poisoning sources will have unwanted consequences. As long as the equation remains profitable, damages will be considered bounds that can be overstepped.

How is it possible to balance the real risk to vulnerable populations if public perception does not consider it a threat and its direct consequences on health are rarely treated in the media?

Preventive campaigns, environmental education, and State regulation can give rise to policies with communication strategies based on solving the potential risk, and the need to take action on the imminent collapse may not arise.

In those cases, crisis communication is an antidote that stops—or attempts to stop—the spread and acts almost as damage control of that which is not included in public discussion. This silence has real effects on citizen health.

The media and their coverage of health, the environment, and the impact of natural resource exploitation influence the perception and public opinion of the risks that these extraction activities entail for communities and the natural and biological environment.

In spite of this, information is framed in a way that usually shifts audiences' attention away from the seriousness of the issues published in order to suggest how they must be analyzed, who the stakeholders are, and which actions can prevent the disaster (Igartua & Muñiz, 2004).

On this basis, it is important to remember the statements of Briggs and Hallin (2016) on health news coverage. The biocommunication proposal is relevant as a social and cultural model to manage health knowledge.

For Briggs and Hallin (2016), media coverage has become one of the most important and visible events of the contemporary world. In the first two decades of the 21st century, constant epidemics and sanitary crises introduced medical experts to the media. In audiovisual networks, it is common for health professionals to be featured in programming. Thus, well-positioned and leading time slots in programming schedules break the logic of mediation and linear broadcast. In this sense, journalism and mediatizations establish frameworks according to which biomedicalization and biocommunicability can understand and interact with each other based on the inclusion of primary sources as direct communicators in media.

Likewise, media coverage of events affecting nature—whether possible or caused by deliberate actions—proposes strong links to the health and the social and economic development of communities, with different forms of risk. Mining and open-pit extractivism are economic practices with a social, sanitary, and ecologic impact on the areas exploited. Communication media break the linear transmission and become frameworks for understanding the phenomenon (Briggs & Hallin, 2016).

Case study

In order to compare public perception and media treatment of environmental risk in Ecuador and Argentina, the authors chose a collection of writings from two digital media (with no printed edition). The fact that both have environmental sections and their understanding of news management in general have made it possible for us to create a stable collection that can, essentially, be compared.

Environmental concerns in Latin America are different from those in countries with a production that destroys habitats and nature. Their tradition as commodity producers, truncated industrialization, and the development of a service economy, with a budding introduction to the knowledge economy, create concerns about contamination, climate change, and ecological disasters considered global information, while regional uncertainty revolves around political and economic stability.

In both cases, the concern is related to information quality, understood as accuracy in "obtaining and confirming the data used, the way in which they are organized, and the clarity with which they are communicated" (Luchessi, 2013, p. 116) as well as in the balance with which voices are presented and how communication is structured.

Plan V (*www.planv.com.ec/*) is a digital medium for investigative journalism from Ecuador founded over ten years ago. It focuses on covering electoral debates, politics, and administrative corruption. In the section entitled "Plan Verde" (Green Plan), it is possible to search the archives from March 2022, which mainly has articles focused on waste management, recycling, large-scale mining, illegal mining, legal actions related to the environment, and oil spills.

The environmental topic of mining is directly related to State contracts or alliances with private—and mainly foreign-capital—companies, and in the elections that were held on 20 August 2023, mining and oil exploitation were part of a national referendum: mining in the province of Pichincha and oil exploitation in Yasuní. Most of the population in the province opposed mining and the same result was obtained nationally as regards keeping oil underground in the Yasuní reserve.

With this background, the coverage and discussion of legal and illegal and industrial or artisanal extraction of metals in this digital medium is significant, as it represents an editorial line against transnational companies with contributions from experts and environmental activists who also represent communities and NGOs.

From a total of 45 publications—from March 2023 up to the closing of this publication, on 31 October 2023—14 of them are about Yasuní and oil exploitation, and nine are related to mining. In both cases, this digital medium features renowned authors, such as environmental activists, indigenous rights activists, and rural communities, as well as former members of political parties.

The "Plan Verde" section has two strong narrative lines: a descriptive, testimonial, and purely journalistic one with a variety of sources and an argumentative line that, through data, numbers, and testimonies by academic and scientific experts, strengthens the position of the authors of each article, shared by the digital medium.

As to the first one, the digital media's works are supported by international environmental journalism networks, such as Earth Journalism Network. In these articles, multimedia use is limited, and mainly animated infographics are used to illustrate economic investments and locate areas at risk of mining collapse or oil spills.

The statements of Briggs and Hallin (2016) about biomediatization are nowhere to be found, despite the fact that argumentative works are prepared by experts in social incidence, not scientists. That is to say that information treatment continues to be linear and expository but with arguments to strengthen the positions against interventions by transnational or Ecuadorian extraction companies.

The use of numbers, with economic and sanitary statistics, both from experts and from journalists, does not support the statements made by Briggs and Hallin (2016) since there is still a mediation of the digital media without considering the social sanitary notion—as asserted by the authors—which means that only risks are reported, but no solutions to these practices are proposed.

The "Plan Verde" section includes plenty of legal data: references to statements by authorities, environmental attorneys, and experts who analyzed consultancy services that applied their recommendations; it also includes plenty of figures that compare the oil and mining industries, labour used, immigration, impact on the economy, and several social indexes on education, development, and health. In both cases, from a social sciences viewpoint only and not from the biomedicalization perspective, scientists and experts on health, mining, and oil exploitation can contribute their knowledge and have conversations with social scientists and ONG activists.

In this regard, biomediatization (Briggs & Hallin, 2016) of the reality of mining and oil exploitation does not apply in this digital medium that, although positioned as an activist and community partner, does not propose frameworks for biomedicalization and biocommunicability to understand and interact with each other, since diseases caused by both industries are the product of State corruption and the unethical actions of companies.

In Argentina, most environmental publications are online and on social media. Led by academics, activists, and some journalists that join them in this purpose, mediatization of environmental issues falls on NGOs, scientific research centres, general information sections of generalist newspapers, and individual cyberactivists.

To focus on risk treatment in the media, the authors will analyze the publications of eldiario.ar (www.eldiarioar.com). The medium was founded in December 2020. Built on the experience of eldiario.es (www.eldiario.es) and founded in 2012, it has a business model that is far removed from corporate logic. The medium expresses its concern for information that is "increasingly subordinated to economic return in traditional media". Thus, it proposes to its audiences the goal of "protecting a way of journalism with quality information as its foundation" (Eldiario.ar, n.d.).

Structured like a generalist medium, eldiario.ar has an Environment section subordinated to the Society section. For the purposes of our analysis, the authors looked at the 168 articles published from 1 March 2023, up to the closing of this publication on 31 October 2023. Most of the articles are descriptive: 112 of the ones analyzed for this work provide information, 14 are interviews with environmental leaders, 26 include analyses, and 16 are opinions. Their main focus is the economic consequences of environmental policies, as well as the consequences of the lack of regulation for the activity that prioritizes profit over public interest.

In this sense, 32 of the texts analyzed are about Government, legislative, and judicial actions related to the environment, and 21 are about the economic impact of public or private decisions on the environment. Climate change is addressed in 28 published articles, followed by contamination (26), environmental disasters (17), and environmental activism (16). The rest deal with concerns for extractivism (6), global political consequences (5), and different topics grouped under the miscellaneous category (17). In the last two months analyzed, a Plants and Gardening subsection was introduced. Even though the topic will supposedly stabilize over time, given the small incidence of the collection examined, the articles on these topics are included in the miscellaneous category.

The idea of risk arises more frequently in analytical and opinion pieces. Some interviewees also invoke—on a smaller scale—the concept of environmental risk, disasters, contamination, or economic collapse due to the consequences of environmental damage. In most cases, the controversy lies in the tension between the possibility of profitability—supported by extractivism, fracking, use of toxic agrochemicals, and policies that enable them—and the idea of prevention, environmental care, and crisis control.

Considering the goals of the medium, most articles are informative pieces based on primary sources and framed in a way that gives the regulatory capability of States and multilateral organizations the chance to regulate climate change, extractivist actions, and contamination and to mitigate the risk on the environment and the health of the ecosystem. Although climate, weather, and environmental phenomena put people at risk, eldiario.ar focuses on the responsibility of institutions and the citizen's capacity to control through activism.

Thus, this accountability (Arato, 2002) balances the State's failure to act, corporate greed, and citizens' ignorance. According to Nadal (2011), in the first decade of the 21st century, there were over a million cyberactivists in Argentina. The media organizes—with them and for them—the information that will be used as raw material to design specific actions and leaves the creation of political guidelines to NGOs, multilateral organizations, and State bodies. However, the section is framed in a way that is aimed specifically at striking a balance between financial opportunities (exploitation of lithium, green hydrogen, oil and fuels, clean energy) and the need to protect climate, plant and animal life, and human health.

The equal treatment of sources does not erase the editorial position defended by eldiario.ar. The environment—and the dividends accrued in the territory—need to be cared for and defended against corporations, extractivism, and environmental damage by those who capitalize on environmental predation. In this case, the statements of Briggs and Hallin (2016) about biomediatization do not apply either. Journalists and experts in social incidents write argumentative storytelling pieces, but no intervention of scientists is detected. The treatment of information is linear and expository but with arguments to strengthen the positions against interventions by transnational or extraction companies.

Except in opinion pieces, mostly authored by journalist Marina Aizen, biomediatization (Briggs & Hallin, 2016) of the reality of mining and oil exploitation does not apply to eldiario.ar. The medium takes the position of a community partner as it reports events that may put communities at risk. However, it does not propose frameworks in which biomedicalization and biocommunicability can understand and interact with each other since diseases caused by the industries it refers to are caused by carelessness in policy handling and corporate greed.

Conclusions

From this perspective, the phenomenon is similar to that of Ecuador. The concepts that Briggs and Hallin developed do not apply insofar as the publications have an interest in the issue (Gomis, 1991), and interpretations of the media do not help explain. In a contextual survey of the generalist press, the construction of environmental events is focused on protests and the impact of actions on the economy. This agenda is not consistent with the problems of sustainable development, climate quality, or environmental protection. Sometimes, articles are included in Economy sections, other times in Politics sections, and yet others in the information related to Science and Technology.

In some cases, the information receives "soft" treatments, such as making some species look exotic or highlighting the extravagant actions of those who manage products that have an impact—both positive or negative—on the environment. The public perceives the environmental situation as plentiful. In view of the scarcity and uncertainty generated by political, economic, and financial handlings, the strength of the land, animals, and the natural ecosystem appear as more stable variables. The danger—which supports crisis situations—in social perception is more frequently linked to the actions of human beings related to transactions, proceedings, and regulations than to the consequences these may have on the land.

In this context, risk leads straight to the construction of crises. Unpredictability becomes the backdrop of most everyday situations, and uncertainty is normalized in social environments in which economic variables are a bigger cause for concern than environmental contamination.

Both media tend to be cautious in their treatment when foreseeing, correcting, and handling danger and the resulting crises. With strong informative offerings, both Plan V and eldiario.ar are very strict about the information published.

As to the way sources are addressed, after ensuring that newsmaking is correct, both tend to present data in an organized manner to enhance understanding. Certainty lies in storytelling, as an organized construction of data, rather than in the consequences of the actions described. The choice of focusing on policies,

Government management, and intervention of multilateral agencies and NGOs is directly in tune with the social interactions that are typical of politics in Latin America. According to Castro:

> The strictest sense in which to state that the environmental aspect of the crisis shows the issues of a global economic structure shaped and managed based on a paradigm that [excludes human beings from the laws of nature] while, at the same time, considers the biosphere as an endless reservoir of resources.
>
> (Castro, 2000, p. 47)

The greed of production increasingly affects the finiteness of natural resources, the quality of the environment, and living conditions. However, crisis perception, as a concept, is linked to political, economic, financial, and social uncertainty. Thus,

> Where LAC countries differ from the rest of the world is in the composition of their emissions. The energy sector, including electricity generation, transportation and the use of fuel in industrial processes, accounts for 43 percent of total CO_2-eq emissions, well below the global average of 74 percent.
>
> (Cárdenas & Orozco, 2022, p. 5)

In any case, contamination caused by farming, the use of agrochemicals, and their resulting impact on population health is incidentally covered in traditional media and in an informative manner in those that are part of this paper.

This focus on the economy and the need for political intervention in order to lead sustainable processes that add resources that create some economic certainty makes the media analyzed establish a scaffolding of information that makes it clear that, while unwanted effects may be mitigated, including industries in the economy has a direct impact on the possibility of certainty.

Anticipating unwanted results for the natural ecosystem seems less upsetting than the inability to anticipate political and economic results. For people living in the south of the Americas, the most complex risk is not related to natural resources—which are not perceived as scarce goods yet—but the risk related to the inability to predict in the middle and long term.

Actions by the media—as a political and cultural ecosystem (Scolari, 2015)— do not contribute to a stable and certain construction of economic, political, environmental, and production needs.

Media moguls, who are associated with other industries, choose to create uncertainties on other topics. The focus on uncertainty is transferred to politics, the economy, and other aspects that jeopardize media leadership on social perception and the insertion of their groups in other industries.

In other contexts, the chosen media focus their publications on information without alarming or avoiding the issues that directly affect daily life. In both, risk and crisis are elsewhere. Uncertainty as an expression of political and economic contexts blurs storytelling about fracking, contamination, and impacts on the health of the population near mining areas.

Although weather shows the changes caused by high emissions of gases, the region suffers its consequences, such as droughts and flooding, greater temperature range, and extinction of species; the highest risk and vulnerability is limited only to the perception of risk arising from socio-economic ecosystems.

Uncertainty—which usually leads to a crisis—clings to the decisions on the economy and society made by other citizens with management responsibilities. Thus, communication rides on two issues: imminent danger of economic restrictions and constant crisis, for which there is no social consensus.

Immersed in these frameworks, the media studied present dissenting voices that balance those that defend the potential profit of general contamination. The univocal nature of perceptions is supported by these multiple voices. And, at the same time, crisis normalization is built on putting risks at the same level. Always in turmoil, always lacking foresight.

References

Arato, A. (2002). Accountability y sociedad civil. In E. Peruzzotti & C. Smulovitz (Eds.), *Controlando la Política. Ciudadanos y Medios en las Nuevas Democracias Latinoamericanas.* (pp. 23–85). Editorial Temas.

Briggs, C. L., & Hallin, D. C. (2016). *Making health public: How news coverage is remaking media, medicine, and contemporary life.* Routledge.

Cárdenas, M., & Orozco, M. (2022). *The challenges of climate mitigation in Latin America and the Caribbean.* United Nations Development Programme (Latin America and the Caribbean Policy Documents Series N°. 40).

Castro, G. (2000). La crisis ambiental y las tareas de la historia en América Latina. *Papeles de Población, 6*(24).

De Almeida, N., Castiel, L., & Ayres, J. R. (2009). Riesgo: concepto básico de la epidemiología. *Salud Colectiva, 5*(3), 323–324.

Ecuador, Ministery of Energy. (2021). *El Gobierno Nacional Expidió el Reglamento General de la Ley Orgánica de Eficiencia Energética.* www.recursosyenergia.gob.ec/el-gobierno-nacional-expidio-el-reglamento-general-de-la-ley-organica-de-eficiencia-energetica/

Eldiario.ar. (n.d.). ¿Quiénes somos?. *Eldiarioar.com.* www.eldiarioar.com/quienes-somos/

Gomis, L. (1991). *Teoría del periodismo. Cómo se forma el presente.* Paidós Ibérica.

Igartua, J. J., & Muñiz, C. (2004). Encuadres noticiosos e inmigración. Un análisis de contenido de la prensa y televisión españolas. *Zer. Revista de Estudios de Comunicación, 9*(16), 87—04. https://doi.org/10.1387/zer.531

Luchessi, L. (2013). ¿Noticia o contenidos? Esa es la cuestión. In L. Luchessi (coord.), *Calidad informativa. escenarios de postcrisis* (pp. 113–121). La Crujía.

Nadal, H. (2011). Testimonio: ciberactivismo y medio ambiente. El caso de Greenpeace Argentina. *Nueva Sociedad, 235.*

Pizarro, R. (2001). *La vulnerabilidad social y sus desafíos: una mirada desde América Latina.* Economic Commission for Latin America and the Caribbean, Statistics and Economic Projections Division.

Presidencia de la Nación Argentina. (2022). *AR guideline 1 "dosimetry factors for external and internal exposure, guideline levels of radionuclides in food and water, and recommendations for radon gas exposure control.* www.argentina.gob.ar/noticias/la-arn-actualizo-la-guia-ar-1-con-recomendaciones-para-cumplir-con-la-norma-basica-de

RAE. (2023). *Diccionario de la lengua española* (23a ed.). https://dle.rae.es/peligro

Riorda, M. (2020). Antes de comunicar el riesgo o la crisis hay que diferenciarlos. *Más poder local, 41*, 20–23.

Scolari, C. (2015). *Ecología de los medios: entornos, evoluciones e interpretaciones.* Gedisa.

9 Media coverage of risk and expert opinion

Noel Pascual-Presa
and Tania Forja-Pena

Introduction

The relationship between the media and science has been the subject of study for decades. The dissemination of health information or information about health risks has gained greater importance and relevance in our lives in recent times, especially considering the recent events related to the COVID-19 pandemic. In this context, the media played a crucial role in how the public perceived and understood health-related risks. On the other hand, the role of scientific experts was also fundamental. Their knowledge, supported by solid evidence, allowed them to provide information and establish guided measures that helped the population make informed decisions that contributed to mitigating potential health risks. However, there are other types of risks that have not received and continue not to receive media attention, while other types of risks have received more coverage. Radon has been one of the greatest overlooked issues in the media agenda for decades, with the responsibility for communicating about radon primarily falling on scientific experts. In this chapter, we will delve into the relationship between the media and scientific experts regarding radon and the communication they carry out about this gas.

Risk communication and information sources

Risk is defined as "things, forces, or circumstances that pose a danger to people" (Stern & Fineberg, 1996, p. 215). While the world today is becoming an increasingly safer place and many of the risks that humans could face have been mitigated, there are others that remain invisible or hidden from a large part of the population. Among them are risks whose effects are not direct or immediately perceptible but manifest themselves after years or even decades. Examples of these risks include radiation, air pollution, genetic engineering, and environmental contaminants, among others (Rossman & Brosius, 2013). In some cases, these types of risks are downplayed or underestimated, and, therefore, public awareness about them is lower compared to other types of risks whose effects

DOI: 10.4324/9781032618180-12

or consequences are noticeable in the short term. In the case of risks associated with radon, it is crucial to understand the nature of this gas, prevention methods, and possible mitigation measures to reduce its consequences. Therefore, communicating this type of information to the population is of vital importance if we want to reduce its impact.

Risk communication is a fundamental pillar for preparing and educating the population about a risk they may face (Winters, 2021). This type of communication encompasses any exchange of information about health or environmental risks among government agencies, media, scientists, public interest groups, individual citizens, and others (Covello et al., 1988). This concept seeks to understand and analyse the decisions and behaviour in modern society in the face of a risk (Renn, 1991). Risk communication is understood as communication that aims to provide information about a risk, offer advice to reduce or mitigate it, and ensure that this knowledge ultimately removes the risk from the individual's immediate environment (Gunn et al., 2021). Risk communication does not imply alarmism, so it is important for intermediaries to convey information without arousing fear or spreading misconceptions among recipients about a topic. This type of communication differs from other forms of scientific communication in that its objectives are, on the one hand, (1) to engage the public in scientific governance and political decision-making and, on the other hand, (2) to reduce society's potential exposure to risks (Bucchi, 2002; Mantovani et al., 2017). In general terms, the goal of risk communication is to explain the uncertainty of a potentially hazardous situation to the entire affected society and to facilitate the exchange of information among experts, media, and citizens to aid in decision-making regarding if people should accept, reduce, or avoid a risk (Leiss, 1994; Leiss, 1996; McComas, 2006; Mantovani et al., 2017).

According to the study by De las Heras-Pedrosa et al. (2020), which involved conducting a sentiment analysis and understanding of emotions during the COVID-19 pandemic in Spain, the population stated that they formed their opinions through the diversity of journalists, perspectives, and news disseminated in mainstream media. The role of the media in risk communication is crucial (Griffin et al., 1999) because they play a significant role in centralising public discourse (Zhao et al., 2019). However, while the media can convey scientific information to society in an easily understandable language, thereby helping to consolidate and expedite information dissemination, they can also have negative or less positive effects (Frewer et al., 2002; Lofstedt, 2003). For example, they can promote the development of fragmented knowledge (Gibbons et al., 1994; Bartlett et al., 2002). The reach of the media and their potential to influence knowledge can pose a danger to the dissemination of risks. This is because the media agenda chooses which topics to cover in more detail and which ones to address more sporadically and without sustained coverage over time. These choices can lead to a distorted perception of risk by recipients (Rossman & Brosius, 2013). Similar conclusions were reached by Ophir (2018) in his analysis of epidemic coverage in American newspapers, where fragmented communication was observed: "articles emphasizing social and economic disruptions tended to lack a discussion of health implications. Articles that focused on medical and health implications, on the other hand, tended not to discuss social implications" (p. 115).

The role of journalists, communicators, and media outlets in general, acting as intermediaries between science and the public, is therefore essential for creating and disseminating a message that is understandable to their audience. They bear the possibility of crafting a discourse that is comprehensible (Mantovani et al., 2017). Furthermore, media outlets also have the duty and responsibility to bring this risk communication into the media agenda. When media coverage of a particular risk increases, placing it in the public debate, individuals exposed to this information will perceive the risk and make critical decisions about how to respond to it (Rowe et al., 2000).

Informing the public about radon is communicating a risk, and at the same time, it is scientific communication. "Over the last few decades, science communication has increasingly been targeted at the general public with the aim of providing socially relevant information and addressing practical and widespread issues" (Mantovani et al., 2017, p. 185). In this way, risk communication and scientific communication help create an informed and proactive society. However, relying solely on a trustworthy source like scientists is not sufficient for effective communication. The media must also develop new abilities to understand the work of scientists and know how to convey that information to society. On the other hand, scientists must receive training in risk communication to understand how and what information to convey (Crick, 2021).

The role of experts during a crisis or in the face of a risk is of special importance since they are often the most reliable and credible source for the affected population (Reynolds et al., 2022). These are determining factors because the pre-existing trust individuals have in the information source will greatly influence the effects that information has on them, both in their behaviour and commitment (Zhong et al., 2021). Despite this, there are other factors that influence individuals' trust levels (Boyd et al., 2019), such as perceived credibility, information quality, or the relationship with the source (Wood et al., 2022). However, generally, information from expert sources like scientists has positive effects on individuals' behaviour and commitment. Moreover, when risks are closer to and have a direct impact on people, trust in these information sources significantly increases (Entradas, 2022).

On the other hand, it is worth noting that we have been witnessing a shift in population trends towards different sources and communication channels (Smith et al., 2021), especially towards social media. During the COVID-19 pandemic, social media became the primary source of information consulted by the population (Reynolds et al., 2022). Nevertheless, it should be emphasised that a significant portion of the population finds traditional media (television, newspapers, internet, radio) more useful than other interpersonal information sources, like expert scientists, when seeking information in situations of risk (Muturi, 2022). That's why the entire responsibility and burden of informing society about the effects and the mitigation measures of a risk should not fall solely on expert sources. However, they should be considered for information dissemination from now on, regardless of the channel or medium used.

Media coverage of radon gas

The public's perception of a risk is greatly influenced by the dissemination and coverage provided by the media. Therefore, it is of great interest to analyse how information about radon gas is communicated. This will allow us to gain insight into the population's perception and awareness of the risk. To do so, it is essential to understand whether radon is part of the media agenda and to comprehend when and how it is addressed. Identifying those responsible for placing it on the agenda is crucial, whether they are political authorities, scientists, the public, or private companies. In this way, we can assess the effectiveness of risk communication management associated with this gas carried out by one of the key players in contemporary communication, the media.

To address these questions, a content analysis was conducted on five representative Spanish news outlets. The sample was based on the cybermedia maps created by the Novos Medios research group in previous R&D projects. In this case, the final sample consists of 415 news articles analysed (87 from *ABC*, 28 from *El Confidencial*, 74 from *El Mundo*, 150 from *El País*, and 76 from *Eldiario.es*) published from 1976 until 31 December 2022.

The sample was constructed by searching for the word "radon" in the search engines of the selected media websites. All pieces retrieved through this filter were reviewed and analysed to ensure they were referring to the gas under study. For the analysis, a data extraction instrument was created, considering different categories that examined the main theme, focus, news values, sources, geographic area, and risk perception. This analysis was carried out between March and May 2022 and was replicated in March 2023.

From this content analysis, several relevant ideas can be extracted. Firstly, the main topics. Most of the news articles have a healthcare focus, meaning they are related to medicine and prevention in some way. A significant number of articles explain the causes and consequences of lung cancer, how it originates, how to prevent it, and possible treatments. Radon is mentioned as one of the causes of this disease in these articles. Many of them are also related to research and serve as dissemination and communication of a new scientific breakthrough or discovery. This is especially true regarding the concentration of radon in enclosed spaces and methods for mitigating or reducing these levels.

Similarly, there is journalistic production on housing, both in global terms and, more specifically, on buildings or areas where air quality is poor and radon is present. Additionally, there are news articles focused on housing legislation at various levels, from regulations promoted by the European Union to the specific legislation of each country or region.

On the other hand, although initially, we related and studied radon as the gas present in buildings and enclosed spaces that can be harmful to health, it also appears in different contexts, for example, as an element used to study the behaviour of volcanoes, earthquakes, and other phenomena. Radon levels vary before a volcanic eruption or an earthquake, so they must be analysed with the aim of finding ways to prevent and anticipate the effects of these natural phenomena that can be dangerous for the public.

There is also room for more varied news, which, in some cases, could be considered misinformation. In some instances, radon is presented as a positive element for health. For example, in some pieces about tourism, the excellent properties of water with high concentrations of this gas are highlighted, something that has been scientifically proven to be more harmful than beneficial to health. In other pieces, clear acts of misinformation may not be committed, but there is also inadequate reporting on the dangers or consequences of prolonged exposure.

Related to the previous point, it is important to consider and also interesting when analysing news articles' publication times because scientific knowledge and legislation have changed over the years analysed. An example of this is that it was not until 1987 when the World Health Organization declared radon as a human carcinogen. In parallel, as scientific knowledge advanced, the first recommendations for maximum indoor radon gas concentration levels began to emerge, considering that humans should not be exposed to avoid potential health risks. Therefore, news articles from the early years analysed may have inaccuracies regarding the characteristics and effects of radon on health. Still, they cannot be understood as misinformation because the context and knowledge were different from what was known at the time of this research.

On the other hand, it is important to highlight that there is a low interest in innovation when it comes to communicating about radon. Most pieces consist of text and a main image. Few of them opt for other types of multimedia content, such as videos, graphics, and infographics. Those news articles that include videos are primarily agency videos, and those featuring graphics and/or infographics are information paid for by external companies. The weight of news articles paid for by companies or advertorials is very low, but it is noteworthy how these news articles have more editing and a more informative presentation with a higher number of images, graphics, or infographics. Two types of paid content are observed: on the one hand, those published purely for informational and brand image enhancement purposes, and on the other hand, those published to promote a product. An example of the first case is the article published by a pharmaceutical company in the newspaper *El País*, where information about the importance of early detection of lung cancer is disseminated (Cañizares, 2022). An example of the second case is the advertorial from an appliance company where, under the pretext of providing information about how air quality affects telecommuting efficiency, an air filtration system is presented (Ec. Brands, 2021).

In general terms, the role of the media in reporting on radon is superficial. Most news lacks long-term follow-up. They are individual pieces disseminated at a given time due to some motivation, which could be a press release from a public institution, a research group, or an event. There are no in-depth analytical pieces on radon, nor is there a tracking of specific cases. This is why they are considered fragmented pieces without continuity over time. There is very limited participation of civil society in these reports; they simply appear as protagonists or sources in specific cases. Thus, it is not the citizens who take the initiative to contact the media to share their reality. In many cases, journalists or media outlets do not seek out civil society sources to supplement their information. When external sources are used to communicate about radon, it is primarily expert sources that are consulted. These are the ones which usually provide insight into news events involving radon.

Radon is primarily a secondary topic mentioned in relation to other news topics at the moment. On the other hand, there is no attempt to silence or omit the reality of radon's consequences, but there is a very limited knowledge of it and a low interest in making this topic part of the media agenda.

Scientific experts and radon communication

To fully understand the dynamics between scientific experts and the media regarding the issue of radon, it is essential to delve deeper into how these actors interact in the dissemination of information and public awareness about the risks associated with this gas. Scientists and experts play a leading role in the radon issue. Nowadays, they are responsible for conducting research on this gas, as well as communication and sometimes even risk management associated with it. Their specialised knowledge provides authority, credibility, and precision in hazard assessment. Therefore, they should be the primary source of information that the media use to inform about radon. However, as mentioned earlier, the information must be processed beforehand to translate technical language into accessible and easily digestible language for the general public. This is crucial if we want to raise awareness and educate people about the risks of radon, enabling them to take preventive or mitigation measures against them.

To address the analysis of this area, we have opted for in-depth interviews with some of the leading figures in the field of radon research in Spain. All the interviewees have years of experience and are the main representatives in the study of this gas, with some of them even having global recognition. In this way, we aim to both showcase their role in the radon issue and gather their opinions on risk communication carried out by the media and the potential challenges they face in this field.

Sometimes, in the face of a risk, the media are hesitant to share it for fear of instilling a certain level of alarm or concern in the population. Interviews with experts shed some light on this issue. All the interviewees admit not feeling concerned about being affected by radon. This might seem counterproductive or contradictory, considering the potential effects of this gas. The key to this lack of concern lies in the fact that they all have an in-depth understanding of the nature of this gas. Therefore, they know whether they can be affected and know how to take action to mitigate potential effects. Moreover, all of them have taken measures in their homes and workplaces to detect the levels of radon to which they could be exposed. Those who have done so and encountered radon levels higher than recommended have been able to take the necessary steps to reduce these levels and, therefore, their risks.

The solution seems simple: if there is a well-informed and educated population, the risk disappears or is substantially minimised if they follow this trend of action. The problem is that we are still far from reaching that point.

Throughout the chapter, the importance of scientific experts as sources of information, and specifically their relationship with the media, has been emphasised. The analysis of media coverage revealed that expert sources were not always used to communicate about radon. This leads to information that is not always precise, rigorous, or relevant data being omitted for the population. Interviews show that some experts have never been contacted by the media to provide information about radon or its dangers. This demonstrates that collaboration between these two central actors of radon communication is not always possible or does not always happen. Perhaps there is a lack of interest on the part of the media in using these types of sources, or perhaps a lack of awareness about the issue and its risks among journalists themselves. These two reasons would lead journalists to believe that there is no need to rely on them to communicate about the risks of radon, and consequently, the presence of this gas continues to be virtually hidden or sidelined from the media agenda.

The experts themselves also indicate the lack of interest or necessity on the part of the media to publish information about radon. They agree that it is necessary to combat this lack of communication actions, which is largely due to the inactivity of the media. What experts demand is not only more communication from the media but also more rigorous information. For this, it is necessary to rely on them to avoid including information that may have negative or counterproductive effects.

Another aspect to mention, which has been detected in previous analyses, is the existence of news articles in the media that discuss new technologies or methods developed by private companies to mitigate radon. Experts caution against this type of news because when there are economic interests involved, the information can be biased, and the rigour of the content can be lost. Some of the population may perceive this type of news as hidden advertising; therefore, the credibility of the information may be undermined even if it is accurate. It is especially important to exercise caution when official sources are involved because it can damage not only the credibility of that specific news article but also the credibility of official sources for future occasions.

The challenges and possible solutions

As we advance in the study of risk communication associated with radon, it is evident that new challenges arise along the way. One of these challenges lies in the complexity of radon risks themselves. Being intangible, invisible, odourless, and tasteless complicates public awareness of it. Furthermore, its effects, as mentioned, are long-term, and it is not until years or decades later that the consequences become noticeable. All of these make it difficult for the population to prioritise this risk or simply consider it important. Therefore, for the population to become aware of the dimensions and magnitude of this risk, they must receive consistent and non-fragmented information about this gas, as observed in the analysis. This is the only way for people to understand this risk and perceive it as part of their immediate reality. People must comprehend its importance and dimensions and act accordingly.

The main challenge we face is the passivity of the media and the difficulty of changing the dynamics they have been following thus far. Not only is there a lack of interest in communicating about radon, but also a lack of theoretical and critical knowledge to report on the topic, as well as poor contrast and inclusion of diversity in sources. This attitude has direct consequences because sporadic and fragmented information is evidently insufficient to raise public awareness of radon. This lack of interest or passivity, observed throughout society, is not driven by hidden interests but is rather understood as a lack of knowledge about the subject and its real consequences.

On the other hand, greater collaboration between the media and scientific experts is necessary to reverse and improve the situation. It is essential that the dialogue between both parties increases so that scientific experts can effectively convey the information they consider necessary through the media to society. In this way, society will acquire the essential knowledge to act and take appropriate measures.

This set of initiatives also benefits the media by increasing their credibility with the public. It is essential that the population trusts them and the information they disseminate to achieve the desired effects. Raising public interest in radon will help to promote a more informed, critical population capable of preventing the adverse health effects of this and many other risks.

The media should turn to primary sources of information (scientific experts) to effectively educate and inform about risks, especially those in which individual behaviour plays a significant role in reducing or mitigating their harmful effects. Therefore, a cooperative relationship between the media and experts is needed. The scientific community and the media must work hand in hand to build an informed, critical society capable of responding to present and future risks and hazards.

Acknowledgments

This chapter is part of the project *Radon in Spain: Public perception, media agenda and risk communication* (RAPAC), financed by the Spanish Nuclear Safety Council [Consejo de Seguridad Nuclear] (SUBV-13/2021). The author, Tania Forja-Pena, holds a predoctoral contract from the *Xunta de Galicia* (ED481A-2023–043).

References

Bartlett, C., Sterne, J., & Egger, M. (2002). What is newsworthy? longitudinal study of the reporting of medical research in two British newspapers. *BMJ*, *325*(7355), 81–84. https://doi.org/10.1136/bmj.325.7355.81

Boyd, A. D., Furgal, C. M., Mayeda, A. M., Jardine, C. G., & Driedger, S. M. (2019). Exploring the role of trust in health risk communication in Nunavik, Canada. *Polar Record*, *55*(4), 235–240. https://doi.org/10.1017/S003224741900010X

Bucchi, M. (2002). Comunicazione pubblica della scienza e situazioni di rischio: Il caso della mucca pazza. In *La scienza negoziata. Scienze biomediche nello spazio pubblico* (pp. 159–198). Il mulino.

Cañizares, F. (2022, November 8). Cáncer de pulmón, la importancia de la detección temprana. *El País*, October 4, 2023. https://elpais.com/sociedad/transformar-hoy-el-manana/2022-11-08/cancer-de-pulmon-la-importancia-de-la-deteccion-temprana.html

Crick, M. J. (2021). The importance of trustworthy sources of scientific information in risk communication with the public. *Journal of Radiation Research*, *62*(Supplement_1), i1–i6. https://doi.org/10.1093/jrr/rraa143

Covello, V. T., Con Winterfeldt, D., & Slovic, P. (1988). Risk communication. In C. C. Travis (Ed.), *Carcinogen risk assessment* (Contemporary Issues in Risk Analysis) (vol. 3, pp. 193–207). Springer. https://doi.org/10.1007/978-1-4684-5484-0_15

De las Heras-Pedrosa, C., Sánchez-Núñez, P., & Peláez, J. I. (2020). Sentiment analysis and emotion understanding during the COVID-19 pandemic in Spain and its impact on digital ecosystems. *International Journal of Environmental Research and Public Health*, *17*(15), 5542. https://doi.org/10.3390/ijerph17155542

EC. Brands (2021, June 28). Cómo puede la calidad del aire interior afectar a tu eficiencia en el trabajo. *El Confidencial*, October 4, 2023. www.elconfidencial.com/tecnologia/2021-06-28/filtro-calidad-aire-eficiencia-trabajo-bra_3113527/

Entradas, M. (2022). In science we trust: The effects of information sources on COVID-19 risk perceptions. *Health Communication*, *37*(14), 1715–1723. https://doi.org/10.1080/10410236.2021.1914915

Frewer, L. J., Miles, S., & Marsh, R. (2002). The media and genetically modified foods: Evidence in support of social amplification of risk. *Risk Analysis*, *22*(4), 701–711. https://doi.org/10.1111/0272-4332.00062

Gibbons, M., Limoges, C., Nowotny, H., Schwartzman, S., Scott, P., & Trow, M. (1994). *The new production of knowledge: The dynamics of science and research in contemporary societies*. Sage Publication.

Griffin, R. J., Dunwoody, S., & Neuwirth, K. (1999). Proposed model of the relationship of risk information seeking and processing to the development of preventive behaviors. *Environmental Research, 80*(2), S230—S245. https://doi.org/10.1006/enrs.1998.3940

Gunn, C. M., Maschke, A., Harris, M., Schoenberger, S. F., Sampath, S., Walley, A. Y., & Bagley, S. M. (2021). Age-based preferences for risk communication in the fentanyl era: 'A lot of people keep seeing other people die and that's not enough for them'. *Addiction (Abingdon, England), 116*(6), 1495–1504. https://doi.org/10.1111/add.15305

Leiss, W. (1994). Review of risk society, towards a new modernity, by U. Beck, M. Ritter, S. Lash & B. Wynne. *The Canadian Journal of Sociology/Cahiers Canadiens de Sociologie, 19*(4), 544–547. https://doi.org/10.2307/3341155

Leiss, W. (1996). Three phases in the evolution of risk communication practice. *The Annals of the American Academy of Political and Social Science, 545*(1), 85–94. https://doi.org/10.1177/0002716296545001009

Lofstedt, R. E. (2003). Science communication and the Swedish acrylamide "alarm". *Journal of Health Communication, 8*(5), 407–432. https://doi.org/10.1080/713852123

Mantovani, C., Crovato, S., Pinto, A., Mascarello, G., Cortelazzo, M., & Ravarotto, L. (2017). Risk communication by health professionals: An analysis of press releases drafted by italian veterinarians. *Veterinaria Italiana, 53*(3), 185–195. https://doi.org/10.12834/VetIt.796.3843.3

McComas, K. A. (2006). Defining moments in risk communication research: 1996–2005. *Journal of Health Communication, 11*(1), 75–91. https://doi.org/10.1080/10810730500461091

Muturi, N. (2022). The influence of information source on COVID-19 vaccine efficacy and motivation for self-protective behavior. *Journal of Health Communication, 27*(4), 241–249. https://doi.org/10.1080/10810730.2022.2096729

Ophir, Y. (2018). Coverage of epidemics in American newspapers through the lens of the crisis and emergency risk communication framework. *Health Security, 16*(3), 147–157. https://doi.org/10.1089/hs.2017.0106

Renn, O. (1991). Risk communication and the social amplification of risk. In R. E. Kasperson & P. J. M. Stallen (Eds.), *Communicating risks to the public. Technology, risk, and society* (vol. 4., pp. 287–324). Springer. https://doi.org/10.1007/978-94-009-1952-5_14

Reynolds, R. M., Weaver, S. R., Nyman, A. L., & Eriksen, M. P. (2022). Trust in COVID-19 information sources and perceived risk among smokers: A nationally representative survey. *PLoS One, 17*(1), e0262097. https://doi.org/10.1371/journal.pone.0262097

Rossmann, C., & Brosius, H. (2013). The perils of risk communication and the role of the mass media. *Bundesgesundheitsblatt, Gesundheitsforschung, Gesundheitsschutz, 56*(1), 118–123. https://doi.org/10.1007/s00103-012-1588-y

Rowe, G., Frewer, L., & Sjöberg, L. (2000). Newspaper reporting of hazards in the UK and sweden. *Public Understanding of Science (Bristol, England), 9*(1), 59–78. https://doi.org/10.1088/0963-6625/9/1/304

Smith, A., Vrbos, D., Alabiso, J., Healy, A., Ramsay, J., & Gallani, B. (2021). Future directions for risk communications at EFSA. *EFSA Journal, 19*(2), e190201-n/a. https://doi.org/10.2903/j.efsa.2021.e190201

Stern, P. C., & Fineberg, H. V. (1996). *Understanding risk: Informing decisions in a democratic society*. National Academy Press.

Winters, M. (2021). *Contagious (mis)communication: The role of risk communication and misinformation in infectious disease outbreaks* [Dissertations & Theses, Europe Full Text: Literature & Language]. https://search.proquest.com/docview/2493869136

Wood, L. M., D'Evelyn, S. M., Errett, N. A., Bostrom, A., Desautel, C., Alvarado, E., Ray, K., & Spector, J. T. (2022). "When people see me, they know me; they trust what

I say": Characterizing the role of trusted sources for smoke risk communication in the okanogan river airshed emphasis area. *BMC Public Health, 22*(1), 2388. https://doi.org/10.1186/s12889-022-14816-z

Zhao, M., Rosoff, H., & John, R. S. (2019). Media disaster reporting effects on Public Risk perception and response to escalating tornado warnings: A natural experiment. *Risk Analysis, 39*(3), 535–552. https://doi.org/10.1111/risa.13205

Zhong, Y., Liu, W., Lee, T., Zhao, H., & Ji, J. (2021). Risk perception, knowledge, information sources and emotional states among COVID-19 patients in Wuhan, China. *Nursing Outlook, 69*(1), 1–21. https://doi.org/10.1016/j.outlook.2020.08.005

10 Conclusions

Deductions on health communication in the current media scene and challenges for the future

José Sixto-García, Sara Pérez-Seijo and Berta García-Orosa

Health communication

Heath communication has become a highly relevant issue in today's communication scene. Communicating in health involves considering the art and techniques to inform, influence, and motivate the public on health issues of social relevance from individual, community, and institutional perspectives (Busse & Godoy, 2016). In fact, the World Health Organisation (WHO) believes that integrated, effective, and coordinated communication is essential to build a better and healthier future for every person on the planet. Therefore, they propose a strategic approach that includes information and counselling for a multitude of health problems (WHO, 2017), including those arising from work environments (Schober et al., 2022).

The current understanding of health communication (Parkinson & Davey, 2023) encourages improving the relationship between health workers and patients (Busse & Godoy, 2016). It also includes other fields and disciplines, such as organisational communication, marketing, sociology, journalism, or media itself, since communication is essential to the nature and practice of science (Sanz-Valero, 2019). Indeed, accessibility, feasibility, credibility, reliability, relevance, time adaptation, and intelligibility are some of the characteristics that the WHO attributes to effective health communication (WHO, 2017). Ensuring the implementation and establishment of health communication policies requires a commitment on the part of public institutions (Myhre et al., 2022) without damage to the proactive activity coming from citizens themselves (Johansson et al., 2021; Paniagua, 2022).

An advanced society cannot and should not remain oblivious to bringing medical knowledge closer to the population. It is only this knowledge that will contribute both to a greater effect of prevention and to the identification of the symptoms or consequences that a disease may cause. It has an impact on improving people's quality of life, although it must also be borne in mind that not all people are interested in health information, and not all are willing to pay attention to this type of content (Link, 2024). In fact, communication should be one

DOI: 10.4324/9781032618180-13

of the elements that facilitate the search for solutions, but sometimes, it does not flow as it should (Laschinger et al., 2004). Also, strictly in the healthcare field, the training and acquisition of professional competencies related to communication, as well as the evaluation of its application and impact throughout the treatment, constitute a guarantee of quality and continuity of care (Francés-Tecles & Camaño-Puig, 2024). In the same way, the relationship based on trust that is established between doctors and patients is particularly significant for the efficacy of messages in healthcare contexts (Alaszewski & Horlick-Jones, 2003).

Health communication is not only about preventing diseases or reporting symptoms or pathologies but also about identifying risks for the population (Plough & Krimsky, 2013). The fundamental objective of risk communication is to save lives so that people get the right to know how to protect their health and that of their environment according to the information received. This has been the case since The Atomic Age when chemical and nuclear industries in the United States opted to minimise the sense of insecurity and anxiety that these industries generated in citizens. However, the growth of risk communication has not declined since then but it has been consolidating as a discipline that addresses the strategic planning of communication in cases of dangers, accidents, or catastrophes. All of this happens so that the people involved can make both preventive and action decisions (Farré, 2005; Idoiaga et al., 2016; Renn, 1991; Valente et al., 2021).

Risk communication also requires a high degree of commitment on the part of public institutions, the scientific community, the medical community, and the media (Covello et al., 1986) to deliver clear and understandable messages (Herovic et al., 2020). However, this has to be done with sufficient forcefulness to capture the attention of the population and awaken the awareness of citizens. In many cases, messages about risks interfere with the broad media agenda without generating practically any type of impact or guarantee of receptivity on the part of the public. People neither identify the risk nor are even aware that an organisation or public administration is providing them with information that has a direct impact on their health. It is clear, for example, that the public does not have the same perception of the risk associated with smoking as with radon.

The use of appropriate media and health communication methods is a determining factor in the success of health communication plans (Latham & Martin, 1977), so it is essential to conceptualise the needs from which this communication comes (Cassata, 1980). Achieving the desired objectives in health communication is complex, as it is about influencing health behaviours and action-oriented decision-making, so a strategic orientation is fundamental (Kreps, 2012). Narrative formats such as videos or comics can influence the effectiveness of messages (Zhou et al., 2023), as well as other digital applications, such as WhatsApp, have proven successful in today's digital health communication (Kadhuluri et al., 2023). A recent study analyses the usefulness of YouTube channels for the spreading of information on radon but diagnoses an underestimation of the functions of this social network and a poor understanding of radon as a real public health problem (Sixto-García et al., 2024). In the coming years, perhaps the metaverse can also help to improve this situation with the creation of immersive spaces (Ahn et al., 2022; Rahaman, 2022) in which citizens can be made aware of dangers and threats in a highly reliable and trustworthy scenario despite being in a simulation.

Thus, an effective health communication strategy (Mehdizadeh-Maraghi & Nemati-Anaraki, 2023) requires consideration of different aspects in order for the message to be transmitted reliably and safely from the source to the different target audiences (Boyd & Furgal, 2022). Consequently, health communication strategies need to consider at least the following elements:

- **Defining the pending message to be communicated.** Before initiating any communication action, the source must be clear about what it wants to communicate. Therefore, the clarity and conciseness of the message will be decisive for the proposed objectives. The message must be shaped according to journalistic criteria, differentiating the lead from complementary information. This way, the information transmitted is clear and sufficiently relevant for the public to take into consideration.
- **Setting objectives.** In health communication strategies, the formulation of objectives is an essential element. No communication action can be successful if the purpose of the messages is not previously defined. When an organisation, institution, or the medical or scientific community decides to send a message to society, there is always an intention and, therefore, a communication objective. Main and secondary objectives can be set, but the priority is that these objectives correspond to specific actions that can be achieved within a specific time frame and that are realistic in terms of the institution's communication capacity or organisational size. If an objective is not delimited in time, it will be meaningless, as it will never be possible to verify whether or not it has been achieved.
- **Segmenting the public.** As important as determining what to communicate is deciding who to communicate it to. The recipients are not a homogeneous group; it is possible to distinguish different targets according to age, gender, purchasing power, or the geographic area where they live. If there are different types of audiences, it is logical to adapt the message to each segment or to decide which of these segments is really the target audience of the message to be conveyed. In health communication, the same message addressed to everyone will always be less effective than a message adapted to different specific audiences.
- **Strategies and implementation.** Once it has been determined what is to be communicated and to whom, it is time to decide how. This is when the configuration of communication strategies and the model for implementing them come into play. When communicating health, one solution may be to

transmit the message through the media (television, radio, written press, etc.) or dissemination media (posters, information boards, brochures, etc.) that are most widely used by the target audience. A greater number of media does not always imply a better reception of the message, but rather, the effectiveness of the communication is conditioned by the suitability of the message, audience, and channel. When configuring a current strategy, digital media such as mobile applications or social networks must be considered. They are spaces that are part of the daily communication routines of millions of people around the world. Therefore, knowing where users interact, it would make sense to transmit the message in their comfort zone. On the other hand, the usefulness of social networks has also been proven in health emergencies (Wang et al., 2022).

- **Measuring feedback.** No health communication strategy would make sense without evaluating its success and effectiveness and checking whether the objectives formulated at its conception were achieved or not. This explains why it is important that objectives are always designed with a time frame so that after a certain period of time, the degree of achievement can be checked. For the measurement of return to be complete, not only quantitative factors such as audience ratings, the number of views of a video, or the number of comments or likes a post receives on social networks should be considered. Qualitative factors are also essential when talking about health. In this type of campaign, the achievement of social purposes, such as the prevention of illnesses or the identification of symptoms, takes precedence over economic parameters. It is the case of other entities that may focus their communication on the sale of a tangible product. In reality, health communication is based on social responsibility actions that ultimately seek to achieve a better society in the short or medium term.

However, in the conceptualisation of current health communication strategies, it is important to consider two important determinants. Firstly, the management of potential crises that may arise during the development. Therefore, it will be necessary to have a contingency plan that responds to unexpected situations that may be serious or worrying for society. Clearly, the objective of crisis communication (Claeys & Opgenhaffen, 2016) in these cases will always be to minimise the impact of what has happened. That the possible negative consequences do not damage the organisation's image or, at least, that the damage is as little as possible.

Secondly, another challenge that health message senders must face is misinformation, which poses a serious threat to advanced societies (Sixto-García et al., 2021). Therefore, its eradication is a constant concern for supranational organisations such as UNESCO or the European Commission (European Commission, 2018, 2020; Ireton & Posett, 2018). Despite these efforts, there is still no definitive solution to this global problem. Health issues are precisely one of the areas where fake news and malicious information tend to flourish. It was the case, for example, during the COVID-19 pandemic (Bonnet & Sellers, 2019; Kim, 2023; Shirish et al., 2021). Moreover, new digital spaces such as social networks enable the massive and viral distribution of this type of content (Hopp, 2022; Leung et al., 2023). It can result in terrible consequences for users who are unable to identify misinformation and who may be guided by advice or practices proposed by unreliable sources.

Effective risk communication: the case of radon gas

From the point of view of health communication, the different chapters in this book have emphasised the importance of disseminating the risk that radon gas represents for society. It has been ranked as the second cause of lung cancer worldwide and the first among non-smokers (García-Talavera & López, 2019). Understood then as a public health problem, effective communication of radon as a risk stands as a fundamental challenge. However, communication efforts to date have been scarce and ineffective, focusing mainly on risk perception (Fisher & Sjöberg, 1990). Nonetheless, they often neglect the need to disseminate existing prevention and mitigation measures and actions.

The perception pointed out by some authors of this monograph about radon risk prevention agencies not adequately communicating their warnings to the public is directly related to the widespread lack of knowledge about the health risks of radon. In this regard, recent studies highlight risk communication as a crucial aspect of public health response (Thomas et al., 2022).

In the absence of effective communication strategies, Bouder et al. (2021) have published *The Potsdam radon communication manifesto*. This is a document in which they offer recommendations based on 50 years of research and practice in both risk communication and health communication. In it, the authors consider that radon risk communication should aim to influence a number of aspects. Firstly, communication should contribute to an informed citizenry so that they understand the risks of radon gas and its main areas of occurrence or concentration, but without unnecessary scaremongering. Secondly, it should empower citizens so that they can make informed decisions to protect themselves. Thirdly, it should build and reinforce trust in the pertinent health and safety authorities. Finally, it should facilitate conflict resolution through the involvement of both those affected and the various stakeholders.

Based on these objectives, Bouder et al. (2021) offer a series of suggestions aimed at enabling nations to prepare effective communication strategies to overcome the difficulties of communicating the public health risk of radon gas: (1) it is essential for governments and communicators addressing radon risk to implement communication programs grounded in scientific evidence; (2) it is necessary to reframe radon, shifting its perception from being merely a natural radioactive gas to being recognised as a form of indoor air pollution; (3) political leaders need to take the initiative and engage with experts and other stakeholders to draw attention to how serious radon gas really is for public health; (4) communication efforts should aim to be coherent, inclusive and consistent, calling for a more systematic coordination between policy areas and levels of government; (5) communication must be maintained consistently over time, keeping the risk on agenda; (6) interactive tools have the potential to improve communication effectiveness and facilitate informed decision-making; (7) specialised training programs need to be created, as "well-trained communicators may become 'ambassadors' and 'multipliers'" (p. 912); (8) and promote research in the social sciences and humanities related to radon.

The need to raise awareness of the problem of radon gas makes it necessary to formulate indicators of good practice aimed at ensuring clear and appropriate communication, as well as to provide citizens with the necessary tools to protect themselves and make informed decisions. The effectiveness of the strategies also depends on the degree of understanding and accessibility of the messages, so the strategies that are formulated must also consider the target audiences:

- **Strengthen scientific dissemination** of the problem that radon gas represents for public health. It requires researchers to acquire communication skills that allow them to convert complex and technical data into information that is understandable to laypeople, that is, to society in general. It is essential to strip away the traditional opacity associated with scientific language in order to attract public interest in this type of information, which is normally not very attractive (Denia, 2020). Also, to take advantage of the trust placed in scientific experts (Breakwell, 2000) to increase the reach of their messages. Bridging the gap between science and society becomes urgent for the sake of effective communication.
- **Increase media coverage** of radon gas, reinforcing the message about its health effects and highlighting existing measures to prevent or mitigate its presence indoors. Moreover, journalism is understood as a public service that acts as an intermediary between reality and citizens, and it must keep the public educated and informed so they can make fully informed decisions. The presence on the media agenda is still scarce, although there is a progressive increase in Spanish regions (Negreira-Rey & Vázquez-Herrero, 2022). The general lack of knowledge among the population about these gas conditions affects the perception of the seriousness of the risk (Khan & Chreim, 2019; Griffin et al., 2004).
- **Adopt an interdisciplinary approach** to the risk communication of radon, which requires involving all interested parties (scientists, the construction sector, public administration, the media, etc.). On the other hand, it is also necessary to take into consideration the actions they have implemented or the results they have achieved. Risk communication needs an integrative and diversified approach in terms of disciplines, as previous studies have pointed out (Agyepong & Liang, 2022; Bouder et al., 2021). Collaboration between

key actors is needed to generate quality information that will really help to strengthen current communication and effectively raise awareness of the public health problem of this gas. As a last resort, the aim is to enable society to understand, perceive, and assess the risks in order to protect itself (Johansson et al., 2021). The latter is in line with the so-called instructional communication, which is key to public risk communication.

- **Geographically segment communication campaigns** according to the incidence of radon gas in different regions or areas of the territory. In order to avoid generating unnecessary alarmism and to ensure communication adapted to the needs of the area, it is advisable to advocate for local segmentation (municipalities, neighbourhoods, etc.) as a strategy when planning dissemination and public awareness initiatives.
- **Use social media as communication channels** to reach wider audiences. Two arguments support this proposition. First, traditional forms of media limit the reach of the messages transmitted. Secondly, studies have found that increasing numbers of people are using online platforms to inform themselves about issues related to health (Masilamani et al., 2020). With all that, to ensure effective communication, it is crucial that information disseminated through social media is perceived as useful and relevant. Moreover, it must be communicated with caution to avoid causing unnecessary confusion and concern among citizens. Again, it is not only a matter of keeping people informed about the health effects of this gas but also raising awareness of the various prevention and mitigation measures available.

Challenges and trends for the coming years

The present work emerges in the context of major changes in communication and in society in general. Through an in-depth analysis of the case of radon gas, notable trends in public health, communication, and risk are explored and compiled in this final chapter to provide a complete and exhaustive view of the issue.

Although radon gas has been in the public eye for more than a century, it has always faced two difficulties from a communication point of view. These are its persistence over time, the absence of crises, and the "natural" label attributed to it, associating it with positive aspects. This last point, in particular, adds a communicative complexity to the risk, as the perception of the natural tends to minimise its potential danger. Today, these inherent challenges of radon gas risk are accentuated by social transformations and the increasing digitalisation of society.

Trends

Through three large sections, the first one dedicated to risk analysis, a second one linked to communication, and a third one related to reception, a multidisciplinary review is carried out that allows not only a transversal analysis of risk but, above all, marks the trends and challenges that will emerge in the coming years. In the fourth wave of digital communication, among others, the consumption of information in networks, disinformation, and the use of Artificial Intelligence brings new possibilities and challenges, but also difficulties in effective communication away from disinformation.

The volume starts from the need for an interdisciplinary approach that affects all citizens (Bouder et al., 2021; McComas, 2006). The first part, which focuses on exploring problems related to radon and the importance of citizen science, emphasises the public health significance of radon gas, where radon exposure refers to the human inhalation of radon gas and its disbandment products in enclosed spaces. This exposure results in ionising radiation that affects the pulmonary tract and increases the risk of lung cancer, making radon the second cause of lung cancer after smoking. In this context, the relevance of the participation of various social actors, including citizens, is highlighted.

In the second part, we find some of the main actors of public opinion with the greatest influence on awareness in the field of health. These are the media, especially the digital natives (Zhao et al., 2019), who are particularly relevant in the field of health and information sources. Several authors allowed us to observe how local communication (Fleming et al., 2006) emerges with great relevance mainly in the field of risks that continue over time.

Digital native media, as channels with structures and techniques naturally adapted to the digital environment, occupy a key position in the current communication ecosystem. Many of the initiatives that have emerged in the last twelve years in the field of public health communication derive from digital native media, which have become an emerging global phenomenon and have contributed to promoting responsible and corroborated communication.

Finally, the work has reviewed the last part of the life of risk, the perception, and consequent adoption of preventive or mitigating measures. Overall, the work confirms the link to risk and the need for good communication (Hong et al., 2019; Rowe et al., 2000; Frewer et al., 2002; De Vries et al., 2021; Dryhurst et al., 2020; Abrams & Greenhawt, 2020; Malecki et al., 2021) and other major trends for the coming years which are discussed further on.

The first major trend is the need for a multidisciplinary approach to the topic, not only because of the liquid and interrelated characteristics of today's society but also because of the idiosyncrasies of radon gas itself. The work has reviewed the current situation of radon gas as a public health concern and also from the point of view of the relevance of effective communication (Rohrmann, 2008) and prevention. The relevance of raising awareness among the population in general and, in particular, among those living or working in the most affected areas of the relevance of good prevention and mitigation of radon gas for public health stands out as the fundamental theme in a volume in which international experts from different fields of knowledge involved in the object of study, from medicine to communication, have participated.

The second major trend to note is the relevance of digitalisation in our societies and, thus, in the treatment of radon risk. In the era where technologies and algorithms influence practically every aspect of daily life, contemporary risks are increasingly mediated (Lupton, 2016), and it is vital to accumulate knowledge about these risks, especially at the social level. Effective communication no longer only involves traditional actors but also includes new ones, such as the public, in co-production processes (Sixto-García & Quintillán-Poza, 2022). Therefore, we have a double objective: to know but also to raise awareness of the situation with radon gas, as one cannot be aware of the dangers if one does not know the risks of exposure (Khan & Chreim, 2019), the different actors, and uncertain situations in the current context. Citizen science, an approach whereby the public collaborates with radon experts during the research process, could help increase not only knowledge but also protective actions. Within this aspect, the relevance not only of experts but, above all, in the processes of co-production of information, already analysed previously in other areas, stands out (Sixto-García & Quintillán-Poza, 2022). Citizen science, an approach whereby the public collaborates with radon experts during the research process, could help increase not only knowledge but also protective actions.

The third trend is the relevance of communication and innovation in the construction and circulation of messages (from narratives to visualisation or relations with audiences). The outbreak of the COVID-19 pandemic was a significant turning point in health communication. In this new outlook, health communication and awareness campaigns have become increasingly relevant. Media professionals have adjusted their approaches and are now more aware of the need for effective and non-sensational communication.

The growing digital audience poses a pressing need to develop innovative and disruptive communication strategies. This challenge is aggravated by the increased segmentation of audiences across diverse platforms, formats, and resources. It is imperative to adapt to changes in digital consumption habits, especially in younger demographic segments, for whom the virtual field represents a new dimension for socialisation and information.

In this context, the metaverse emerges as a non-mass alternative, particularly effective when calls to action actively engage these audiences with the content presented. Media coverage and the involvement of local and international experts become crucial to ensure the effectiveness of these communication strategies.

Digital media can actually be more effective than traditional media in amplifying information about health risks (Ng et al., 2018).

Finally, researchers in the fields of risk and crisis management have increasingly come to conceptualise risk and crisis as cross-border (for instance, Boin et al. 2014; Falkheimer & Heide, 2015). This means that the origins and effects of hazards and adverse events cross-functional aspects and national and cultural boundaries. This coordinated work between authors from different universities, fields of knowledge, and countries has confirmed the trend.

Challenges

Risk communication faces many challenges in the coming years. There are many challenges and opportunities from the point of view of research, prevention, and communication, starting with redefining frameworks, instruments of analysis, and challenges. However, the challenges also include supranational strategies involving not only experts and communicators but also public administrations and citizens. Legislators, political parties, public administrations, lobbies, and educators will shape public health risks in the coming decades through their actions.

Radon gas is described as a problem that goes through different phases and requires different interventions, especially in the most affected geographical areas. Throughout the three parts, the work reflects on the current situation but, above all, provides solutions and identifies uncertainties for the near future. Without wishing to be exhaustive, some of the challenges from the point of view of research, teaching, and political praxis are proposed in what follows.

Research

- **International holistic approach.** Holistic analyses of radon gas are proposed, considering both its local aspects and its international dimension. A longitudinal and multidisciplinary approach is advocated to understand the complexity of this phenomenon. Research should transcend borders and collaborate at a global level.
- **Responsible and non-alarmist communication.** A critical reflection on the importance of honest, continuous, and non-alarmist communication of radon gas by all actors involved is proposed. This would mean balancing risk disclosure with the promotion of solutions and preventive measures, thus encouraging an informed and balanced perception in society and contextualised and useful information.
- **Methodological innovation.** New methodologies to address the radon gas problem are actively sought. Research should explore innovative approaches and advanced technologies that allow for more efficient detection, accurate measurements, and more effective mitigation strategies.
- **Strategies**. An understanding of the strategies, information flows, and interrelationships between the various actors involved in the radon gas issue is essential. This in-depth knowledge will enable the coordination of efforts and the development of effective strategies at local, national, and international levels.
- **Ongoing studies.** Research should be continuous, interdisciplinary, and adaptive. Emphasis is placed on the need for ongoing studies to keep up with the changing dynamics of radon gas knowledge in different areas and communication, allowing strategies and policies to be adjusted in response to new evidence and findings.

Expert praxis

- **Experts as disseminators.** The need to work closely with experts who act as sources of information for the media can be drawn from the work. Collaboration between the scientific community and the media, although complicated by different codes and work processes, will contribute to accurate and understandable reporting on radon gas.
- **Effective and specialised communication.** Clear communication strategies by experts in different areas related to radon gas are advocated. Dissemination of specialised information will facilitate public understanding and strengthen confidence in the proposed preventive measures while creating a bond of trust with the recipients. It is important to promote clear communication strategies between the centres of expertise in the different fields involved.
- **Co-production with the citizenry.** Need to involve the public in the elaboration of messages. The importance of citizen participation is recognised, and the need to involve the public in the elaboration of messages is highlighted. The inclusion of the community in the creation of communication strategies ensures the relevance and effectiveness of the proposed measures.

Public administration praxis

- **Raising awareness.** The need to raise awareness in the public sphere of the importance of keeping the public well informed, usually through expert sources but with informative language, is highlighted. Transparency and open communication are essential. This prior work will be essential in the face of a potential crisis.
- **Coordination between institutions.** As in the case of experts, the importance of clear communication strategies from public administrations and between different expert areas is highlighted. Collaboration between government bodies and experts ensures a coordinated and coherent response.
- **Citizenry involvement.** Active citizenry participation in message development is crucial. The public administration could facilitate spaces for citizen participation, promoting an inclusive dialogue that strengthens the effectiveness of communication strategies.

Media praxis

- **Accurate information.** Quality information on radon gas in the media is essential. The responsibility of the media to provide accurate facts, avoid the generation of unnecessary alarms, and contribute to an informed understanding by the public is emphasised. The involvement of specialised journalists is recommended.
- **Visibility of preventive measures.** Visibility and explanation of knowledge about prevention and mitigation measures through the media are a main point.

Acknowledgments

This chapter is part of the project *Radon in Spain: Public perception, media agenda and risk communication* (RAPAC), financed by the Spanish Nuclear Safety Council (SUBV-13/2021), and the R&D project *Digital-native media in Spain: Strategies, competencies, social involvement and (re)definition of practices in journalistic production and diffusion* (PID2021–122534OB-C21), funded by MICIU/AEI/10.13039/501100011033 and by "ERDF/EU".

References

Abrams, E. M., & Greenhawt, M. (2020). Risk communication during COVID-19. *Journal of Allergy and Clinical Immunology: In Practice*, *8*(6), 1791–1794. https://doi.org/10.1016/j.jaip.2020.04.012

Agyepong, L., & Liang, X. (2022). Mapping the knowledge frontiers of public risk communication in disaster risk management. *Journal of Risk Research* (online first). https://doi.org/10.1080/13669877.2022.2127851

Ahn, S. J., Kim, J., & Kim, J. (2022). The bifold triadic relationships framework: A theoretical primer for advertising research in the metaverse. *Journal of Advertising*, *51*(5), 592–607. https://doi.org/10.1080/00913367.2022.2111729

Alaszewski, A., & Horlick-Jones, T. (2003). How can doctor's communication about risk more effectively? *BMJ*, *327*, 721–731. https://doi.org/10.1136/bmj.327.7417.728

Boin, A., Rhinard, M., & Ekengren, M. (2014). Managing transboundary crises: The emergence of European union capacity. *Journal of Contingencies and Crisis Management*, *22*(3), 131–142. https://doi.org/10.1111/1468-5973.12052

Bonnet, J. L., & Sellers, S. (2019). The COVID-19 misinformation challenge: An asynchronous approach to information literacy. *Internet Reference Services Quarterly*, *24*(1–2), 1–8. https://doi.org/10.1080/10875301.2020.1861161

Bouder, F., Perko, T., Lofstedt, R., Renn, O., Rossmann, C., Hevey, D., Siegrist, M., Ringer, W., Pölzl-Viol, C., Dowdall, A., Fojtíková, I., Barazza, F., Hoffmann, B., Lutz, A., Hurst, S., & Reifenhäuser, C. (2021). The potsdam radon communication manifesto. *Journal of Risk Research*, *24*(7), 909–912. https://doi.org/10.1080/13669877.2019.1691858

Boyd, A., & Furgal, C. (2022). Towards a participatory approach to risk communication: The case of contaminants and Inuit health. *Journal of Risk Research*, *25*(7), 892–910. https://doi.org/10.1080/13669877.2022.2061035

Breakwell, G. M. (2000). Risk communication: Factors affecting impact. *British Medical Bulletin*, *56*(1), 110–120. https://doi.org/10.1258/0007142001902824

Busse, P., & Godoy, S. (2016). Comunicación y salud. *Cuadernos.info*, *38*, 10–13. https://hdl.handle.net/20.500.12724/2339

Cassata, D. M. (1980). Health communication theory and research: A definitional overview. *Annals of the International Communication Association*, *4*(1), 583–589. https://doi.org/10.1080/23808985.1980.11923826

Claeys, A. -S., & Opgenhaffen, M. (2016). Why practitioners do (not) apply crisis communication theory in practice. *Journal of Public Relations Research*, *28*(5–6), 232–247. http://doi.org/10.1080/1062726X.2016.1261703

Covello, V., von Winterfeldt, D., & Slovic, P. (1986). Communicating scientific information about health and environmental risks: Problems and opportunities from a social and behavioral perspective. In V. Covello, A. Moghissi, & V. R. R. Uppuluri (Eds.), *Uncertainties in risk assessment and risk management*. Plenum Press.

De Vries, M., Claassen, L., te Wierik, M., Das, E., Mennen, M., Timen, A., & Timmermans, D. (2021). The role of the media in the amplification of a contested health risk: Rubber granulate on sport fields. *Risk Analysis*, *41*(11), 1987–2002. https://doi.org/10.1111/risa.13731

Denia, E. (2020). The impact of science communication on Twitter: The case of Neil deGrasse Tyson. *Comunicar*, *65*, 21–30. https://doi.org/10.3916/C65-2020-02

Dryhurst, S., Schneider, C. R., Kerr, J., Freeman, A. L. J., Recchia, G., van der Bles, A. M., Spiegelhalter, D., & van der Linden, S. (2020). Risk perceptions of COVID-19 around the world. *Journal of Risk Research*, *23*(7–8), 994–1006. https://doi.org/10.1080/13669877.2020.1758193

European Commission. (2018). *A multi-dimensional approach to disinformation: Report of the independent high level group on fake news and online disinformation*. Publications Office of the European Union.

European Commission. (2020). *Fighting disinformation*. https://ec.europa.eu/info/live-work-travel-eu/health/coronavirus-response/fighting-disinformation

Falkheimer, J., & Heide, M. (2015). Trust and brand recovery campaigns in crisis: Findus Nordic and the horsemeat scandal. *International Journal of Strategic Communication*, *9*(2), 134–147. https://doi.org/10.1080/1553118X.2015.1008636

Farré, J. (2005). Comunicación de riesgo y espirales del miedo. *Comunicación y Sociedad*, *3*, 95–119.

Fisher, A., & Sjöberg, L. (1990). Radon risks: People's perceptions and reactions. In S. Majundar, R. Schmalz, & E. Miller (Eds.), *Environmental radon: Occurrence, control and health hazards* (pp. 398–411). The Pennsylvania Academy of Science.

Fleming, K., Thorson, E., & Zhang, Y. (2006). Going beyond exposure to local news media: An information-processing examination of public perceptions of food safety. *Journal of Health Communication*, *11*(8), 789–806. https://doi.org/10.1080/10810730600959705

Francés-Tecles, E., & Camaño-Puig, R. (2024). Experience and perceptions of nurses regarding the communication process with patients. *Revista de Comunicación y Salud*, *14*, 1–15. https://doi.org/10.35669/rcys.2024.14.e342

Frewer, L. J., Miles, S., & Marsh, R. (2002). The media and genetically modified foods: Evidence in support of social amplification of risk. *Risk Analysis*, *22*(4), 701–711. https://doi.org/10.1111/0272-4332.00062

García-Talavera, M., & López, F. J. (2019). *Cartografía del potencial de radón en España*. Consejo de Seguridad Nuclear. https://bit.ly/3UXMvRV

Griffin, R. J., Neuwirth, K., Dunwoody, S., & Giese, J. (2004). Information sufficiency and risk communication. *Media Psychology*, *6*(1), 23–61. https://doi.org/10.1207/s1532785xmep0601_2

Herovic, E., Sellnow, T., & Sellnow, D. (2020). Challenges and opportunities for pre-crisis emergency risk communication: Lessons learned from the earthquake community. *Journal of Risk Research*, *23*(3), 349–364. https://doi.org/10.1080/13669877.2019.1569097

Hong, Y., Kim, J. S., & Xiong, L. (2019). Media exposure and individuals' emergency preparedness behaviors for coping with natural and human-made disasters. *Journal of Environmental Psychology*, *63*, 82–91. https://doi.org/10.1016/j.jenvp.2019.04.005

Hopp, T. (2022). Fake news self-efficacy, fake news identification, and content sharing on Facebook. *Journal of Information Technology & Politics*, *19*(2), 229–252. https://doi.org/10.1080/19331681.2021.1962778

Idoiaga, N., Gil De Montes, L., & Valencia, J. -F. (2016). Communication and representation of risk in health crises: The influence of framing and group identity. *International Journal of Social Psychology*, *31*(1), 59–74. https://doi.org/10.1080/02134748.2015.1101313

Ireton, C., & Posett, J. (Eds.). (2018). *Journalism, 'fake news' and disinformation: Handbook for journalism education and training.* Unesco.

Johansson, B., Lane, D., Sellnow, D., & Sellnow, T. (2021). No heat, no electricity, no water, oh no!: An IDEA model experiment in instructional risk communication. *Journal of Risk Research, 24*(12), 1576–1588. https://doi.org/10.1080/13669877.2021.1894468

Kadhuluri, D., Hense, S., Kodali, P. B., & Thankappan, K. R. (2023). How WhatsApp is transforming health communication among frontline health workers: A mixed-method study among midwives in India. *Journal of Communication in Healthcare, 16*(3), 268–278. http://doi.org/10.1080/17538068.2023.2189376

Khan, S. M., & Chreim, S. (2019). Residents' perceptions of radon health risks: A qualitative study. *BMC Public Health, 19*(1), 1114. https://doi.org/10.1186/s12889-019-7449-y

Kim, M. (2023). A direct and indirect effect of third-person perception of COVID-19 fake news on support for fake news regulations on social media: Investigating the role of negative emotions and political views. *Mass Communication and Society* (online first). https://doi.org/10.1080/15205436.2023.2227601

Kreps, G. L. (2012). Translating health communication research into practice: The importance of implementing and sustaining evidence-based health communication interventions. *Atlantic Journal of Communication, 20*(1), 5–15. http://doi.org/10.1080/15456870.2012.637024

Laschinger, H. K. S., Finegan, J. E., Shamian, J., & Wilk, P. (2004). A longitudinal analysis of the impact of workplace empowerment on work satisfaction. *Journal of Organizational Behavior, 25*(4), 527–545. https://doi.org/10.1002/job.256

Latham, C., & Martin, G. M. (1977). Rural health communication. *Annals of the International Communication Association, 1*(1), 569–577. http://doi.org/10.1080/23808985.1977.11923707

Leung, J., Vatsalan, D., & Arachchilage, N. (2023). Feature analysis of fake news: Improving fake news detection in social media. *Journal of Cyber Security Technology, 7*(4), 224–241. https://doi.org/10.1080/23742917.2023.2237206

Link, L. (2024). Health information as "fodder for fears": A qualitative analysis of types and determinants of the nonuse of health information. *Health Communication* (online first). https://doi.org/10.1080/10410236.2024.2312611

Lupton, D. (2016). Digital risk society. In A. Burgess, A. Alemano & J. Zinn (Eds.), *Routledge handbook of risk studies* (pp. 301–309). Routledge.

Malecki, K. M., Keating, J. A., & Safdar, N. (2021). Crisis communication and public perception of COVID-19 risk in the era of social media. *Clinical Infectious Diseases, 72*(4), 697–702. https://doi.org/10.1093/cid/ciaa758

Masilamani, V., Sriram, A., & Rozario, A. (2020). eHealth literacy of late adolescents: Credibility and quality of health information through smartphones in India. *Comunicar, 64*, 85–95. https://doi.org/10.3916/C64-2020-08

McComas, K. A. (2006). Defining moments in risk communication research: 1996–2005. *Journal of Health Communication, 11*(1), 75–91. https://doi.org/10.1080/10810730500461091

Mehdizadeh-Maraghi, R., & Nemati-Anaraki, L. (2023). Application of problematic integration theory in health communication: A scoping review. *Health Communication* (online first). http://doi.org/10.1080/10410236.2023.2281078

Myhre, S. L., French, S. D., & Bergh, A. (2022). National public health institutes: A scoping review. *Global Public Health, 17*(6), 1055–1072. https://doi.org/10.1080/17441692.2021.1910966

Negreira-Rey, M., & Vázquez-Herrero, J. (2022). La cobertura mediática sobre el gas radón en los medios digitales en Galicia. *Prisma Social, 39*, 4–24.

Ng, Y. J., Yang, Z. J., & Vishwanath, A. (2018). To fear or not to fear? Applying the social amplification of risk framework on two environmental health risks in Singapore. *Journal of Risk Research, 21*(12), 1487–1501. https://doi.org/10.1080/1366987 7.2017.1313762

Paniagua, P. (2022). Elinor Ostrom and public health. *Economy and Society, 51*(2), 211–234. https://doi.org/10.1080/03085147.2022.2028973

Parkinson, J., & Davey, J. (2023) The importance of health marketing and a research agenda. *Health Marketing Quarterly, 40*(4), 347–351. https://doi.org/10.1080/07359 683.2024.2271780

Plough, A., & Krimsky, S. (2013). The emergence of risk communication studies: Social and political context. In T. S. Glickman & M. Gough (Eds.), *Readings in risk*. RFF Press.

Rahaman, T. (2022). Into the Metaverse—Perspectives on a New Reality. *Medical Reference Services Quarterly, 41*(3), 330–337. https://doi.org/10.1080/02763869.2022. 2096341

Renn, O. (1991). Risk communication and the social amplification of risk. In R. E. Kasperson & P. J. M. Stallen (Eds.), *Communicating risks to the public. Technology, risk, and society* (p. 4). Springer. https://doi.org/10.1007/978-94-009-1952-5_14

Rohrmann, B. (2008). Risk perception, risk attitude, risk communication, risk management: A conceptual appraisal. In *15th International Emergency Management Society (TIEMS) Annual Conference*. https://bit.ly/4aJokh1

Rowe, G., Frewer, L., & Sjöberg, L. (2000). Newspaper reporting of hazards in the UK and Sweden. *Public Understanding of Science, 9*(1), 59–78. https://doi. org/10.1088/0963-6625/9/1/304

Sanz-Valero, J. (2019). Comunicación para la salud laboral. *Medicina y Seguridad del Trabajo, 65*(256), 173–176. https://doi.org/10.4321/S0465-546X2019000300001

Schober, D. J., Carlberg-Racich, S., & Dirkes, J. (2022). Developing the public health workforce through community-based fieldwork. *Journal of Prevention & Intervention in the Community, 50*(1), 1–7. https://doi.org/10.1080/10852352.2021.1915736

Shirish, A., Srivastava, S., & Chandra, S. (2021). Impact of mobile connectivity and freedom on fake news propensity during the COVID-19 pandemic: A cross-country empirical examination. *European Journal of Information Systems, 30*(3), 322–341. https:// doi.org/10.1080/0960085X.2021.1886614

Sixto-García, J., García-Orosa, B., González-Lois, E., & Pascual-Presa, N. (2024). Transparency on YouTube for radon risk communication. *Revista Latina de Comunicación Social, 83*, 01–20. www.doi.org/10.4185/RLCS-2024-2266

Sixto-García, J., & Quintillán-Poza, A. (2022). La cocreación como técnica para aumentar el valor de los productos periodísticos. *Estudios sobre el Mensaje Periodístico 28*(3), 703–712.https://doi.org/10.5209/esmp.78544

Sixto-García, J., Rodríguez-Vázquez, A. I., & López-García, X. (2021). Verification systems in digital native media and audience involvement in the fight against disinformation in the Iberian model. *Revista de Comunicación de la SEECI, 54*, 41–61 http://doi. org/10.15198/seeci.2021.54.e738

Thomas, M., Kaufman, S., Klemm, C., & Hutchins, B. (2022). The co-evolution of government risk communication practice and context for environmental health emergencies. *Journal of Risk Research* (online first). https://doi.org/10.1080/13669877.2022 .2077414

Valente J. -P., Gouveia, C., Neves, M. -C., Vasques, T., & Bernardo, F. (2021). Small town, big risks: Natural, cultural and social risk perception. *PsyEcology, 12*(1), 76–98. https:// doi.org/10.1080/21711976.2020.1853946

Wang, H., Xiong, L., Wang, C., & Chen, N. (2022). Understanding Chinese mobile social media users' communication behaviors during public health emergencies. *Journal of Risk Research, 25*(7), 874–891. https://doi.org/10.1080/13669877.2022.2049621

WHO. (2017). *Strategic communications framework.* https://bit.ly/3bVBZDM

Zhao, M., Rosoff, H., & John, R. S. (2019). Media disaster reporting effects on public risk perception and response to escalating Tornado warnings: A natural experiment. *Risk Analysis, 39*(3), 535–552. https://doi.org/10.1111/risa.13205

Zhou, M., Ramírez, S. A., Chittamuru, D., Schillinger, D., & Ha, S. (2023). Testing the effectiveness of narrative messages using critical health communication, *Journal of Communication in Healthcare, 16*(2), 139–146. http://doi.org/10.1080/17538068.2023.2189363

Index